KANSAS
FORTS & BASES

KANSAS
FORTS & BASES

SENTINELS ON THE PRAIRIE

DEBRA GOODRICH BISEL &
MICHELLE M. MARTIN

THE
History
PRESS

Published by The History Press
Charleston, SC 29403
www.historypress.net

Front cover, top middle: Collection of Debra Goodrich Bisel; *bottom middle*: United States Army; *back cover, bottom left*: Buckner family. Unless otherwise noted, all images appear courtesy of Michelle M. Martin.

First published 2013

Manufactured in the United States

ISBN 978.1.60949.826.9

Library of Congress CIP data applied for.

CONTENTS

FOREWORD

Many Americans today sadly but all too typically think of Kansas as merely "that state in 'the middle' where they grow wheat." Such a simplistic perception unfortunately robs them of knowledge of the rich, varied and fascinating history of a part of this country that played a vital role in the nation's westward expansion and development and that continues to host institutions that are integral to the American way of life. Thankfully, *Kansas Forts and Bases: Sentinels on the Prairie*, by authors Deb Goodrich Bisel and Michelle Martin, presents a very welcome giant stride forward in correcting that egregious misperception.

This superb "must read" book is not simply a catalogue listing of historically significant locations; it is a much-needed and extensively researched historical account of Kansas from the mid-eighteenth century, when Native Americans and French fur traders roamed its hills and prairies, to the twenty-first century, where America's oldest active military post west of Washington, D.C., today hosts the intellectual center of the U.S. Army.

This comprehensive account of the founding and fate of Kansas' historic forts and bases is brought to vivid life with numerous eyewitness stories taken from the letters, journals and memoirs of the men and women who actually *lived* the history you'll find in these pages. The authors have crafted a compelling read by weaving into their narrative the *real* stories of the *real* people.

The book's "cast of characters" reads like a "who's who" of American history: explorers Lewis and Clark; fiery abolitionist John Brown; notorious Confederate partisan raider William Quantrill; Civil War soldiers J.E.B. Stuart, George Armstrong Custer and William Tecumseh Sherman; frontiersmen

Buffalo Bill Cody and Wild Bill Hickok; World War II generals Dwight D. Eisenhower, Douglas MacArthur and George S. Patton; and today's military notables, such as generals Colin Powell and Don Holder. The military units covered in this book that made history at Kansas' forts and bases include some of America's most famous ones: First Regiment of Dragoons; Seventh U.S. Cavalry Regiment; African American "Buffalo Soldiers"; and the U.S. Army's "Big Red One" First Infantry Division.

The fires of my own interest in this fascinating subject were stoked during three tours of duty at one of the most historically significant of Kansas' forts—Fort Leavenworth, where I spent one-fifth of my entire thirty-six-year military career, more time than at any other posting. Sitting on the bluffs overlooking the Missouri River where the Santa Fe Trail crossed the river, Fort Leavenworth's history dates back to 1827 when it was the then half-century-old United States' westernmost outpost. The fort was a vital bastion and rallying point during the era of America's westward expansion, the 1846–1848 War with Mexico, the 1854–1856 Border War between "Free State" Kansas Jayhawkers and proslavery Missouri Bushwhackers, the Civil War and the Indian Wars. Fort Leavenworth's Command and General Staff College, established by General Sherman in 1881, educated the U.S. Army officers who led the country's efforts in World War I, World War II and the Cold War, a critical role it continues to admirably fulfill in the twenty-first century. History at Fort Leavenworth is a palpable presence. During my final tour of duty there, I had the privilege of living for five years in the most historic building on post—"The Rookery," dating from 1832, is the oldest building in Kansas and in 1854–55 was occupied by Kansas' first Territorial governor, Andrew Reeder. The fort's founder, Colonel Henry Leavenworth, ordered the building constructed between 1832 and 1834 to serve as his quarters and office, and its frame of hand-hewn timbers and its thick, sturdy native rock walls have endured for nearly two centuries. Young Lieutenant Douglas MacArthur once lived in "The Rookery." In 1867, George and Libbie Custer lived next door. The ghosts of Kansas' early pioneers still haunt "The Rookery's" rooms and hallways—no kidding!

Yet, Fort Leavenworth's story is only *one* of those vividly described and thoroughly recounted in the pages of *Kansas Forts and Bases*. Immerse yourself in the pages of this book, and you'll truly experience real history brought to life.

—Colonel (Ret.) Jerry D. Morelock, PhD
Former Director of Combat Studies Institute, Fort Leavenworth
Editor in Chief, *Armchair General* magazine

ACKNOWLEDGEMENTS

N o book is written solely by its authors. Years of research, the influence of mentors and the support of countless individuals feed an author's soul and provide the nourishment needed to produce a book. We are indebted to so many people that have assisted, guided, cajoled, poked and prodded us along our path that resulted in the book you now hold in your hands. Before we thank any of them, however, our greatest debt of gratitude is to the men and women of the United States military and their families. Without them, we would have no stories to tell on the pages of this volume. The land that became Kansas deserves much praise and thanks. As transplants to Kansas, we both fell in love with the land, people and intricate history of Kansas. The story of Kansas is the story of America and the development of the West and its impact on our imagination and vision of what it means to be an American.

I must thank the following individuals for their support, guidance and friendship that made this book possible—my family: Dale, Susan, Michael, Amelia and Nicholas Martin; my grandmothers: Janet Kerbs, Winnifred Martin and Madalene Martin; my grandfathers: Francis Martin, Roy Kerbs and Michael Bowen; Bill Kurtis, native Kansan, broadcast journalist and friend; Mary Kristin Kurtis, my good friend; Arnold Schofield, Rosemary Frey, William Fischer and Galen Ewing from Fort Scott National Historic Site; Alan Chilton and Deborah Wood from Wilson's Creek National Battlefield; George Elmore from Fort Larned National Historic Site; Tim Rues from Constitution Hall State Historic

Site; Dr. Jeff Broome, Carolyn Cooper, Kendall Gott, Lawana Leach, Kip Lindberg, Kay Little, Matt Matthews, Dr. John Monnett, Douglas Scott and the late Dr. Peter Schmitt, all mentors and friends. Special thanks to my good friend Jenna Blum, an authoress and inspiration! Special thanks to Dr. Donald Fixico for his friendship, support, advice and encouragement. My greatest thanks and love to Debra Goodrich Bisel for her friendship, creative partnership and love of all things Kansas. We may not be sisters by blood, but our sisterhood has been forged by triumph, tragedy, loss, success and much laughter! I couldn't have asked for a better friend or co-author.

—Michelle Martin

My first thanks must go to my husband, Gary, who supports me in all my endeavors, who follows me out the door with my phone or the notes I need, who makes my life full. My circle of friends—creative, supportive, a little bit nuts—believe in me, and I love them for it. I must also thank my sister, Denise, who got the math genes in the family and keeps up with messes I make.

In historical circles, there are too many people to list here that deserve thanks and recognition, but I must name a few: the Civil War Round Table of Eastern Kansas; the Shawnee County Historical Society and especially Doug Wallace, who waits by the phone to give me information; likewise, George Laughead in Ford County who is on constant stand-by; Tom Perry, chronicler of all things J.E.B. Stuart and generous friend; the Heritage Center in Dodge City; Barton County Historical Society; Lecompton Historical Society; and Tim Rues; and Bill Buckner for sharing his family's photos and stories.

My most profound thanks must go to Fort Leavenworth, especially to the Command and General Staff College and Combat Studies Institute. I am privileged to have many friends at this post and to have participated in media training, staff rides and other functions that have given me the opportunity to spend quality time at this most historic fort. I spent time in Bell Hall, now long gone and replaced by Lewis and Clark, with hallways as long as football fields. The students who pass through the Command and General Staff College are the best in the world, and so are their instructors starting with D.K. Clark, Ed Kennedy, Steve Tennant, Tom Chychota, Bob Burns, Scott Weaver, Scott Porter, Tom Meara, John Pilloni, Bill Lambert, John Barbee, Charles Heller, Steve Boylan, Bill McCollum, Steve Kerrick, Bill Raymond and especially Ken Gott at CSI—a very incomplete list! Also, my "brother," Dave Chuber, historian at Fort Leonard Wood, Missouri, who retired from the army at Fort Leavenworth and is a constant source of information and inspiration.

A few years ago, I was participating in a media exercise with John McWethy, former ABC newsman turned consultant. He asked me to deliver the "after action" report to the retired generals who had led the exercise (including Lieutenant General Don Holder, who is included in this book); McWethy had to visit with the post commandant, General Bill Caldwell. Later, we met in the MacArthur room of the Lewis and Clark Hall to compare notes. When I opened the door, there was John on his laptop at that huge desk with the specter of Mac looming over him. I laughed out loud, and so did he. We then sank into those huge sofas, and he asked why I loved this post so much. "I love the energy, the urgency," I told him. "The energy is palpable. Many of these instructors have children in service, children who will serve under the officers they are training. That lends an incredibly personal touch to their mission. No other place has such an urgent quest for knowledge and understanding."

John and I visited for a long time that day; I could not get enough of his stories of combat and the Pentagon, wrangling with Rumsfeld and endless war stories of a conscientious journalist. We parted company and planned for the next media exercise just weeks away.

That was the last time I saw John. He was killed in a skiing accident on February 6, 2008. There was an e-mail in my mailbox, sent from him just a day or so before he died, notes on issues he wanted to cover in the coming visit to Fort Leavenworth. His loss was devastating, not only to friends and family, but also to the community of soldiers and reporters, a community that needed his insight and expertise. I am forever grateful to him. He left a good example for us to follow, for the roles of reporter and historian are interchangeable; they bear the same profound responsibilities.

I can think of no one better suited to share this job than my dear friend, Michelle Martin. She possesses amazing energy and ability, and no one has a greater passion for Kansas history. No one has a greater passion for sharing it. I am proud to call her my friend and partner in crime.

—Deb Bisel

Acknowledgements

We express our profound thanks to Colonel (retired) Jerry Morelock for providing the foreword to this book. He could have written the entire volume off the top of his head.

Special thanks to Becky Lejeune and the team at The History Press for their hard work. They help historians like ourselves preserve and share our amazing history.

INTRODUCTION

Blessed are the peacemakers…
—Matthew 5:9

The relationship between Kansas and the science of war is so ingrained, so consistent and so evident, yet it seems antithetical to the quiet, conservative, independent farmer that is the quintessential image of the state. It is not. The same values created both, and both created Kansas.

The first Kansans, Native Americans of various tribes, were an independent lot, living well off the land and protective of it. The next Kansans, Spaniards and Frenchmen, lived among the Indians, fit well into their world for the most part and were an independent lot as well.

The next wave of Kansans, Americans, came to Kansas for economic opportunity and to separate themselves from the world of the east and its restraints. In Kansas, a man could breathe free. Literally, they hoped, all men would be able to breathe freely in the new state.

All this independence has come with a cost—defending it. No state knows better than Kansas the price of freedom. It has been paying for more than a century.

Soldiers. Sailors. Flyers. Kansas has raised them, trained them and deployed them. The forts and bases of this state have literally been central to America's defense and remain crucial today. While not every post is given equal time in this volume, it is sincerely hoped that the reader will gain a greater appreciation for the diverse story of Kansas in that common quest to make the neighborhood and the nation safe.

Only with security can one enjoy the quiet life of a farmer.

Chapter 1

POLICING THE INDIAN FRONTIER AND GUARDING KANSAS TERRITORY

1744–1861

How do I like my new home? you say—pretty well—we are happy very happy all to each other...our time is mostly passed in reading...and then when our books become irksome, we ride, fish, and walk, collect all the pretty flowers we see and try to become botanists, the flowers here far surpass those of Leavenworth in fragrance.
—letter from Charlotte Augusta Swords (wife of Captain Thomas Swords) to Lieutenant Abraham R. Johnston, Fort Scott, August 7, 1843[1]

With the stroke of a pen, President Thomas Jefferson doubled the size of the young American nation when he purchased the Louisiana Territory from Napoleon Bonaparte in 1803. Exceeding his presidential powers, Jefferson's vision of the West and its importance to the future development of America fueled his controversial decision. By 1830, settlers had been moving west for decades. The settlement of the West was well underway when President Andrew Jackson championed the Indian Removal Act in 1830. Removing the Five Civilized Tribes from the Southeast, Jackson saw the untapped west as the perfect place to relocate the tribes. Little did he know his actions opened Pandora's box on the plains and prairies. In response to the movement of peoples, a line of military posts was developed to safeguard settlers and the developing trails westward and keep peace among the Native American nations in the West. Trade was also a paramount function of many of these early posts, as they became safe havens and gathering places for traders. These initial sentinels on the prairie were the first in a long line of fortifications that would play a critical role in the history of the American West and Kansas. Life for the first military families and soldiers was filled with hardship, danger, excitement, privation and loss. Their experiences blazed the trail future generations would follow in Kansas.

FORT ATKINSON: 1850–1854

Lieutenant Colonel Edwin Vose Sumner was no stranger to army life, its rigors and the dangers that came with duty. A career soldier, Sumner was appointed to the First U.S. Dragoons immediately after its creation by Congress on March 4, 1833. He would travel west and serve with distinction in the War with Mexico at the Battle of Molina del Ray as a part of Colonel Stephen W. Kearney's Army of the Southwest. Upon the conclusion of the war, Sumner was rewarded for his service and promoted to the rank of lieutenant colonel of the First Dragoon regiment and was stationed in New Mexico. Along with his military duties, he was appointed governor of New Mexico Territory from 1851 to 1853. His intimate role in the shaping of New Mexico would lead to his steadfast belief that the trail network was paramount to settlement in the West. Trade and commerce flourished with the cessation of hostilities, and the Santa Fe Trail became a superhighway of the West. Guarding the trail became one of the paramount functions of the army in the West.

The Santa Fe Trail was one of many transportation arteries that the army depended on to move men and materials in the West. With the opening of the gold fields in California, the Santa Fe and California Trails would become more important and guarding them fell to the men of the army. To protect the trail, in what would become Kansas Territory, Sumner established Fort Atkinson (also known as Fort Sod, Fort Sodom and Fort Sumner) on the left bank of the Arkansas River in 1850.

Originally named Fort/Camp Mackay in honor of Aeneas Mackay (a deputy quartermaster general), the fort was rough-hewn and barren. On the Kansas prairie, sod was plentiful; trees were not. Soldiers cut the tough earth into bricks and stacked them to build the fort. Charles Hallock, a visitor to Fort Atkinson in 1852, noted, "The scarcity of fuel and grass is the chief inconvenience experienced by this fort, though in other respects it is by no means agreeably situated."[2] Fort Atkinson had but one purpose—to guard the lucrative trade that passed along the Santa Fe

General Simon Bolivar Bucker and his wife before his posting to Fort Atkinson, Kansas Territory, while serving in the U.S. Army. *Courtesy the Buckner Family.*

Trail. Its location near the Arkansas River and the trail made it a gathering place for Native American tribes as early as 1850 when the United States government began encouraging the tribes to negotiate peace treaties with the federal government. Hallock noted that lands surrounding the fort were "a barren waste land, without vegetation, save a few shrub bushes and the crispy buffalo grass…the fort itself is of adobe roofed with canvass, containing fair accommodations for the garrison, and defended by a few small field piece and the usual armament."[3]

After a short time, Sumner ordered that the post be deconstructed and rebuilt a short distance away. Soldiers referred to the post as Fort Sod and Fort Sodom because of its building material. By 1854, the post was all but abandoned, as its distance from Fort Leavenworth made resupply difficult. To ensure that the post did not fall into the hands of the Cheyenne or Kiowa, it was destroyed in October 1854.[1]

Sumner would never serve at Fort Atkinson. Upon his arrival in Kansas Territory, Sumner was charged with pacifying the factions warring over the expansion of slavery into the developing territory. With the First U.S. Dragoons as his right arm, Sumner not only had to contend with the three competing political groups operating in Kansas but also the potential threat of attack from Native American tribes in the western regions of the territory. By 1857, Sumner commanded campaigns against the Cheyenne and left the territory in 1858. Today, sadly, nothing remains of the first post founded in Western Kansas to guard the Santa Fe Trail. A gleaming white marker stands on the approximate location of the post three miles west of Dodge City.

FORT BAIN: 1857–1858

The pen is truly mightier than the sword. On May 30, 1854, President Franklin Pierce picked up his pen and signed the Kansas Nebraska Act into law, setting off a series of events that would lead America to war. White settlement would now be permitted in the former Indian lands of Kansas Territory. With the floodgates open, settlers from the North and South streamed into the territory. Settlers would decide through popular sovereignty the fate of the new state—free or slave. The era known as Bleeding Kansas was born, and all eyes watched as Kansans fought one another over the issue of slavery.

Into this maelstrom of political discontent came a man imbued with a sense of righteous anger and indignation toward slavery and its proponents—John Brown. After arriving in Kansas Territory in 1854, Brown quickly established antislavery strongholds from which he, his sons and supporters could operate against the proslavery forces in Kansas. Fort Bain, also called Fort Bourbon, was nothing more than a log cabin built by Brown with assistance from Captain Bain. Their goal was to protect antislavery advocates from the depredations of proslavery forces from Missouri. Fort Bain's location in Bourbon County close to the Missouri state line made it important for gathering intelligence information, assisting the transport of runaway slaves into Kansas and northward to Canada and served as a place for Brown's supporters to resupply while on the move.

Once established, Fort Bain served as a place for antislavery advocates, including James Montgomery, to gather and plan their activities in Southeastern Kansas. On December 2, 1857, Brown and his supporters held off a force of close to five hundred proslavery men in an attack on the fort, incurring a loss of only four men. The fort was less important as a defensive location. John Brown used the fort as a place to retreat, regroup and plan his December 1858 raid into Missouri to free slaves and move them to freedom in Canada along the Underground Railroad. After the violence of the 1850s ceased, the cabin became a private home and today noting remains of Fort Bain.

BROWN'S STATION: 1854–1859

The men were freezing, near starved and shaking with fever. When he stepped down from his wagon seat, John Brown stood in amazement at the sight that greeted his eyes. His sons huddled in crude, half-finished cabins lingering near death in what would become present-day rural Franklin County. With zeal and vigor, Brown assumed control of Brown's Station, and soon discipline and order reigned. With his arrival in Kansas Territory on October 7, 1855, at the behest of his sons, John Brown infused the antislavery forces in Kansas with new energy. Brown's Station would be the launch point for Brown's activities in the territory and play an important role in one of the most infamous and inflammatory events in the tumultuous year of 1856—the Pottawatomie Creek Massacre.

While never a true military post, Brown's Station was home to the Pottawatomie Rifles, a local antislavery militia unit that boasted John Brown Jr. as a member, along with his other siblings who had made Kansas their home. From Brown's Station, Old Osawatomie Brown would write letters back east to inform his supporters about the progress of the war against slavery in the territory.

> *Browns Station, Kansas Territory, 14th Decem 1855*
> *Orson Day Esqr White Hall NY*
> *Dear Sir*
> *I have just returned for the Kansas War (about which you have no doubt learned by the news papers;) & find your Letter of the 19th Nov. As I intend to send you shortly a paper published here giving you a more full account of the invasion than I can consistently afford the time to give; I will only say at this time that the territory is now entirely in the power of the Free State men; & notwithstanding this result has been secured by means of some bravery, & tact; with a good deal of trickery on the one side; & of cowardice, folly, & drunkeness on the other yet so it is; & I believe the Missourians will give up all further hope of making Kansas a Slave State.[5]*

On the night of May 24, 1856, John Brown and his men would leave Brown's Station and use the cover of darkness to conceal their movements to strike at what they believed to be the heart of proslavery activity in the region—Pottawatomie and Mosquito Creeks. Several proslavery and Southern-sympathizing families lived along the creeks. Brown intended to send a message to the proslavery forces in Kansas and Missouri to make them think twice before attacking Kansas Territory again. Emboldened with zeal and anger after the caning of Senator Charles Sumner on the floor of the United States Senate by Preston Brooks of South Carolina, and as a reaction to the most recent raid against Lawrence, Brown decided the time to strike was at hand.

Coming upon the modest cabin of James Pleasant Doyle and his family, Brown's men forced their way into the home and took Doyle and his sons—

William, Drury and John—prisoner. Claiming to be with the army of the North and wanting information, the men seized the Doyles and proceeded to leave. Mahala Doyle, the wife and mother of the captive men, fell to her knees and begged that the life of her youngest son, John, be spared. The boy was left. Moments later, Mahala Doyle heard a chilling sound—a gunshot. "These men were armed with pistols and large knives…They first took my husband out of the house, then they took two of my sons—the two oldest ones, William and Drury—out…In a short time afterwards I heard the report of pistols…I heard two reports, after which I heard moaning, as if a person was dying; then I heard a wild whoop."[6] Mahala Doyle's long night did not end with the first light of day. The task of locating her dead and missing family members fell upon her shoulders. "My husband and two boys; my sons, did not come back any more. I went out next morning in search of them, and found my husband and William, my son, lying dead in the road near together, about two hundred yards from the house. My other son I did not see any more until the day he was buried. I was so much overcome that I went to the house. They were buried the next day. On the day of the burying I saw the dead body of Drury."[7] Fear compelled her to take her remaining children and leave Kansas Territory, never to return. Brown's men also visited Allen Wilkinson and his wife, Louisa, that fateful night. After killing Wilkinson and leaving his bedridden wife to fend for herself, the men continued to Dutch Henry's Crossing along Pottawatomie Creek in search of the Sherman brothers. Such was the nature of life in Kansas in the 1850s.

By 1858, John Brown, feeling his work in Kansas complete, left Brown's Station, never to return. Taking the war against slavery to the east, he planned and executed a bold raid against the federal arsenal at Harpers Ferry, West Virginia. After his defeat and capture on October 10, 1859, Brown was tried and convicted of treason. Before his hanging on December 2, 1859, Brown scrawled his final prophecy on a piece of paper in his cell: "I John Brown am now quite certain that the crimes of this guilty land will never be purged away, but with Blood. I had…vainly flattered myself that without very much bloodshed, it might be done."[8] Ironically, as his cell was cleaned, a letter was found in the trash. Written by the grieving widow and mother Mahala Doyle, who suffered the loss of her husband and two sons at the hands of Brown and his men on Pottawatomie Creek, Kansas Territory in 1856, the letter indicates that her son John Doyle would be present at Brown's hanging. The train carrying Doyle was held up by a landslide inbound from Tennessee. He would not see John Brown hang "between heaven and earth."[9]

FORT DE CAVAGNIAL: 1744–1764

In the beginning, there were furs.

Everyone valued furs—from the Kansa Indians to the courts of France. In a time before synthetic and waterproof fabrics, furs and pelts were highly prized and became legal tender. They were traded for the necessities and the luxuries of life, for horses and whiskey, for pots and guns. In the eighteenth century, most of the vast interior of the North American continent was home to an abundance of wildlife. An active trade grew between native tribes and the Spanish empire's satellite of Santa Fe and the French settlements at New Orleans and in Canada. The Spaniards and the French crisscrossed what would become Kansas. The journey for the French was primarily by boat, as they traveled the Mississippi and then began exploring its tributaries. The first white men known to have seen the Missouri River were, in fact, French explorers in 1673.

It was primarily the French who expanded the fur trade and military outposts that became necessary for protecting, as well as conducting, commerce. In 1724, Etienne de Veniard, sieur de Bourgmont crossed the Missouri River into Kansas, near the Salt Creek Valley (near modern-day Fort Leavenworth). Bourgmont had a colorful and somewhat controversial military career. After deserting his post at Fort Pontchartrain (Detroit), he was to be arrested, but he made himself too valuable to be imprisoned. Instead, the daring Frenchman was rewarded for his outstanding service to France, and that service was all about getting furs. Bourgmont lived among Indians and apparently traded fairly with them. As respect and trust grew for Bourgmont among the native tribes, so did his knowledge of them and the geography of North America. He was credited for the French-Indian alliance that had resulted in the massacre of a Spanish expedition into what is now Nebraska, earning him the gratitude of the French crown. In 1720, he was commissioned as a captain in the French colonial army and named "Commandant of the Missouri River." His orders from the crown were to establish a fort on the Missouri River and to establish peace among the tribes, as well as the sometimes unruly French-Canadian voyageurs, in order to establish a lucrative and steady fur trade. In 1723, Bourgmont established Fort d'Orleans along the Missouri River, midway into what is now the state of Missouri. Fort d'Orleans was the staging ground for Bourgmont's excursions into Kansas.

By 1724, Bourgmont already had a relationship with the Kansa Indians. He had visited their village in 1714 and laid the foundation for a relationship that would last for decades and would establish this area as a cornerstone of westward expansion. In 1725, Bourgmont took chiefs from the Illinois, Missouri, Osage and Oto tribes, as well as the daughter of the Missouri chief, to France. The delegation visited the King's chateaux at Versailles, Marly and Fountainebleau. They hunted with Louis XV. The natives returned home, and Bourgmont was given a noble title. In his absence, Fort d'Orleans fell into disrepair, and as it did, the situation between the Indians and their French neighbors deteriorated as well. By 1729, Fort d'Orleans was abandoned. During the winter of 1732–33, natives killed eleven Frenchmen. It became impossible to keep tabs on the voyageurs, and in 1740, the traders themselves were so unruly that the Indians demanded something be done.[10]

Fort de Cavagnial was authorized in 1743 and built in 1744 on "rising ground overlooking the village of the Kansa Indians on Salt Creek."[11] The exact location remains a mystery. When Merriwether Lewis and William Clark visited the site on July 2, 1803, the foundations and chimneys were still visible.

Today, a marker at Fort Leavenworth suggests the site of Fort de Cavagnial is in view.

According to historian Robert H. Berlin, "The fort was constructed of stout piles, eighty feet on the square, with bastions at each corner. Inside the fort were a commandant's house, a guardhouse, a powder house—all two stories tall—a trader's house and a house for employees of the traders. These were log buildings, most covered with mud, and even the chimneys were made of mud-covered logs. The post's garrison consisted of a commandant, eight to ten soldiers, and several traders. If the French wives of a few soldiers, the Indian wives of the traders, and the children were counted, the post's population was close to forty."[12]

As remote as this location would have been, it would be a mistake, asserted historian Charles E. Hoffhaus,

> …to picture Cavagnial as an isolated post in the midst of hostile aborigines in a vast unexplored wilderness…The Missouri, Kansa, and Pawnee had been allies of the French since early in the 1700's, and voyageurs in sizable numbers had long been trading with them, intermarrying with them, and exploring their territory, starting even before the turn of the century. These tribes used French firearms, tools, cloth goods and cooking utensils. They had come to rely so much on these trade goods that their occasional short supply was a serious hardship and a cause for complaint and the French, on the other hand, avidly sought the furs, which the tribes collected. Many

were converted, or at least exposed, to Christianity…Except for wintertime, the Missouri was a veritable highway at the very door of the fort, carrying traffic at least as far as the Platte and possibly beyond. All four corners of Kansas had been permeated by well-organized expeditions…The Great Chiefs had unfurled the fleur-de-lis of France over at least three principal Indian nations of the area…Bourgmont's efforts had effected a major treaty…allowing free access by French traders clear across Kansas and on to Santa Fe, and this privilege was actively pursued by the traders. Thus, the fort can be placed in context as a control point astride a well-developed trade route characterized, for the most part, by mutual trust and advantage on the part of French and Indians alike.[13]

While the location of this outpost has figured prominently in the defense of America, in 1755 its warriors were fighting *against* the colonists in the French and Indian War. Kansa and Osage were recruited in the area and "participated in the final stages of the celebrated defeat of British Gen. Edward Braddock…The Kansas contingent arrived just after the fight had ended. It is not without the realm of possibility, however, that these recruits from Fort de Cavagnial may have sent a ball or two in the direction of the future father of our country." George Washington, leading the Virginia militia, had taken over when Braddock was mortally wounded. He was in the thick of the rear guard action that held off pursuing Indians. Young Daniel Boone was a wagoner and may have dodged a bullet or two himself.[11]

Though the French lost the war, there was little change in the daily life along the Missouri River until 1764 when the French issued orders to vacate the outposts of the upper Louisiana. The territory was ceded back and forth between France and Spain until finally acquired by the United States in the Louisiana Purchase in 1803.

FORT FLOYD: 1857

Fort Floyd was much like its namesake—a big name and not much else. In fact, it might be argued that this earthen works "fort" had a greater impact on American history than the secretary of war for whom it was named. John B. Floyd was born in Virginia, near what is now the campus of Virginia Tech. His father was an incredibly capable and accomplished man who

Artist Michael Boss painted this depiction of Fort Floyd, Kansas Territory, from historic descriptions. *Courtesy Dr. James Jefferson Broome.*

served in the House of Delegates and as governor of the commonwealth. The son would likewise serve as governor of Virginia and as secretary of war to President James Buchanan. He proved a poor administrator, and later, when an officer in the Confederate army during the Civil War, he would prove a poor soldier as well. (It was Floyd who would surrender Fort Donnelson, Tennessee, to the Yankees.)

Perhaps Lieutenant J.E.B. Stuart, also a Virginian, was aware of Floyd's shortcomings, but he was more aware of his position and his ability to secure promotions. Maybe there was a little brown-nosing when the ambitious officer dubbed a dirt wall about five feet high "Fort Floyd." If Secretary Floyd was impressed when he saw dispatches with references to a "fort" named in his honor, it did not benefit Stuart. In January 1861, Stuart wrote to his brother, Alex, "Floyd has resigned without giving me a captaincy or ordering me to West Point—the latter he certainly had it entirely in his power to do."[15]

While his naming of the fort may not have benefited Stuart, the time spent there would put him in the center of one of the nation's most critical events. In fact, the very fight that found him in this remote part of the Kansas Territory was in itself a pivotal moment in the Plains Indian Wars and would have an effect on the army's perspective for decades to come.

Serving with Colonel Edwin V. Sumner at Fort Leavenworth had already placed Stuart in the thick of Bleeding Kansas. He had been with Sumner following the Battle of Black Jack in June 1856, when they had entered the camp of John Brown and released his prisoners (including Henry Clay Pate, who would die at Yellow Tavern, Virginia, near the spot where Stuart would be mortally wounded in 1864). The year had seen one conflict after another with the sacking of Lawrence, the Battle of Black Jack, the dispersal of the Free State Legislature in Topeka (where the Civil War nearly started) and the Pottawatomie Massacre. By year's end however, it seemed that the Territorial factions were at peace so that attention could be turned to the Indians. "Being no longer occupied with the affairs in this Territory which have caused so much uneasiness, undivided attention can be paid to punishing the Cheyenne Indians," declared General Persifor F. Smith, commanding the Department of the West.[16] Reports had been increasing of attacks on settlers in the western Kansas and Nebraska Territories. Colonel Sumner was ordered to lead a punitive expedition against the Cheyenne.

Private Robert Morris Peck recalled his commander: "Although then [1857] well advanced in years, with hair and beard white as now, he was still quite vigorous, every inch a soldier, straight as an arrow, and could ride like a Cheyenne. Sumner was the ideal veteran commander, and was idolized by his men. He was a natural-born soldier, and always seemed happiest when there was a fight in sight."[17]

He may have been more idolized by Peck than by Stuart. Sumner obviously had great confidence in Stuart; he had come to rely on Stuart's managerial skills as quartermaster and adjutant, but when Stuart refused to accept yet a third responsibility, he and Sumner were at odds and would remain so. Their regional differences may have affected their relationship as well since Sumner was from New England. Nonetheless, when the three hundred soldiers in Sumner's command came upon about five hundred Cheyenne warriors, Peck and Stuart were in agreement that Sumner's order to "Draw—saber!" was not the best plan.

"It was my intention & I believe that of most company commanders to give a carbine volley & then charge with drawn pistols, & use the sabre as a 'derniere resort,'" wrote Stuart to his wife the next day. "But much to my surprise the col ordered draw Sabre charge when the Indians were within gunshot. We set up a terrific yell which scattered the Cheyennes in a disorderly flight & we therefore kept up pursuit."[18]

Peck had thought the order to be "a serious mistake" but noted that the Indians checked their charge. "The sight of so much cool steel seemed to cool their ardor," he wrote.[19]

The Cheyenne were taken aback by the saber charge. A medicine man, perhaps White Bull, had instructed the warriors to bathe in a certain pool of water that would protect them against the enemy's bullets. No mention had been made of swords. The reaction of the Indians in this fight would lead the military to assume Native Americans would cower in the face of their well-trained military. Many lives would be lost proving just how wrong that assumption was.

The fight along the north fork of the Solomon River became a running battle with the cavalry following the Cheyenne for seven miles, exchanging volleys along the way. Two soldiers and about thirty Cheyenne were killed, including a warrior who shot Stuart in the chest. Stuart recalled:

> *I received a severe wound in the center of my breast from a pistol in the hands of an Indian whilst rushing between him and a brother officer in imminent peril which left me disabled for several weeks…The officer rescued had dismounted to fire at this Indian, then standing, with greater accuracy—but his pistol hung fire and in this predicament the Indian rushed at him, and I dashed between with my sabre giving him a severe head cut—but his pistol was discharged not two feet from my body, covering my face and person with powder. The ball has never been extracted.*[20]

It was that wound and those suffered by eleven other fellow soldiers that necessitated the construction of Fort Floyd. Sumner determined to continue his pursuit of the Cheyenne, and the wounded would be left to recuperate with a company of infantry to guard them. "[W]e turned to and threw up a sod-and-dirt wall about five feet high, enclosing a square plot of probably about fifty feet each way—large enough to contain the little garrison and their animals," wrote Peck.[21]

It was while Stuart was lounging on the prairie in the summer of 1857 that he considered the saber charge and determined that the saber should be drawn more quickly and smoothly and set about designing a means of attaching the weapon that would accomplish that end. The War Department was interested in his device, and he was in Washington, D.C., in 1859 to meet with officials when word came of problems at Harpers Ferry, Virginia. He accompanied Colonel Robert E. Lee, as his aide, to put down the troubles and encountered John Brown for the second time in his life. Stuart described the incident to his mother just weeks later:

The insurgents had retired to and barricaded themselves in the fire engine house with several of the citizens of Virginia prisoners. It was raining slightly. In less than an hour Colonel Lee had reconnoitered the place, relieved the militia guard stationed around the Armory by a guard of Marines, and determined upon his plan of operations, which were successfully carried out early in the morning. The Colonel (I was constantly at his side) told me all his plans at the time, and his reasons for them...I was deputed by Colonel Lee to read to the leader, then called Smith, a demand to surrender immediately and I was instructed after his refusal, which he expected, to leave the door and wave my cap at which signal he had directed the storming party to rush up and batter open the doors and capture the insurgents at the point of the bayonet...I approached the door, in the presence of 2,000 spectators, and told Mr. Smith I had a communication for him from Colonel Lee. He opened the door about four inches (a sort of armistice prevailed at this time) and placed his body against the crack with a cocked carbine in his hand; (hence his remark subsequently after his capture that he could have wiped me out like a mosquito). The parley was a long one. He presented his propositions in every possible shape with admirable tact, but all, amounting to the only condition he would surrender upon to be allowed to escape with his party...I omitted what really I think was the greatest service I rendered the State. When Smith first came to the door I immediately recognized Old Osawatomie Brown who had given us so much trouble out here in Kansas. And not until he was knocked down and dragged out pretending to be dead, and I proclaimed it—did that vast multitude and the world find out that John Smith was Old Brown. No one else could have done that there but myself. I got his bowie knife from his person while he was "possuming," and have it yet.[22]

Brown was executed on December 2, 1859. Neither Stuart nor Sumner would survive the Civil War that they had been warming up for in the Kansas Territory. Private Peck later expressed interest in finding the location of that short-lived but significant fortification. Historians continue to search for traces of the fort's existence.

FORT LANE: 1856–1857

Indiana native James Henry Lane came to Kansas Territory and was an active political partisan on behalf of the free-state cause. Lane became one of Kansas' first senators. *Courtesy Deb Bisel, author's collection.*

Brash, opportunistic, bold, treacherous, dangerous, madman, visionary—all these describe one of the most enigmatic and puzzling figures of nineteenth-century Kansas politics: James Henry Lane. A Hoosier from the state of Indiana, Lane came to Kansas and allied himself with the antislavery forces at work in the territory. Quickly he distinguished himself as an orator, military leader and zealot. His brash approach to Kansas political and military matters made him a target for proslavery forces and, in some cases, free state and antislavery partisans as well.

Amos Lawrence and members of the New England Emigrant Aid Company settled the city of Lawrence as early as 1854. Early on the settlers there realized they had a tactical advantage militarily if they occupied Mount Oread, the city's highest point. This gave nervous New Englanders a vantage point that allowed them to see in all directions across the Kansas prairie. Atop Mount Oread they built Fort Lane, named for James Lane. Used for a year, the stronghold was utilized by free state men to protect the city of Lawrence. Brinton W. Woodward, a veteran of the campaigns to protect Lawrence in the territorial period recalled of the fort in 1898:

The fort on Mount Oread had been located and built, under the direction of Lane, at the point of the bluff coming north, where it drops down to the rather lower level or ridge on which Gov. Robinson's house had stood, and where the first university building [since called North College] was afterward placed. Its site has scarcely even yet been wholly obliterated by grading, and it was directly west [south] of where Mr. Frank A. Bailey's residence now stands. It occupied a sightly and commanding position;…was of irregular outline, following the curve or point of the bluff on two sides, with a straight chord subtending on the south. It was laid up as a loose, dry wall from the rough stone gathered about, to the height of from three to four feet, thus making a show of outline fairly exhibited to the east.[23]

In 1863, during Confederate partisan William Clark Quantrill's raid on Lawrence on August 10, the fort was burned atop Mount Oread as horrified residents watched. Today, nothing remains of Fort Lane.

FORT LARNED: 1859–1878

Writing in 1881, Colonel Henry Inman described the days of the Santa Fe Trail with reverence and awe. "The commerce of the Great Plains over that broad path through the wilderness—the Santa Fe Trail—was at its height, and immense trains rolled day after day toward the blue hills which guard the portals of New Mexico."[24] One can paint a mental picture from his vivid and eloquent words of life on the Santa Fe Trail in Kansas. "Oxen, mules, and sometimes horses, tugged wearily, week after week, through the monotony of their long journey, and their precious freight ever tempting wily nomads to plunder, dissimulation, and murder."[25] Commerce was the lifeblood of Kansas in the nineteenth century. From 1822 to 1843, Santa Fe Trail commerce averaged over $130,000 annually.[26] By 1859, however, trade along the Santa Fe had sharply risen to $10,000,000 annually. From March 1 to July 31, 1859, 2,300 men, 1,970 wagons, 840 horses, 4,000 mules, 15,000 oxen and 1,900 tons of freight departed from Missouri headed for New Mexico.[27] Fortunes built upon trade were tenuous; threats posed by trail robbers, Native Americans and Mother Nature all conspired against the trail-weary teamster taking goods from Kansas City to Santa Fe. Protecting commerce had long been a directive of the army in the west. Forts Leavenworth, Atkinson, Mann and Riley all provided protection for commerce along the trails in Kansas. Small in size, Forts Atkinson and Mann were hastily constructed and short-lived in nature. Only Forts Leavenworth and Riley garrisoned adequate troop strength to ensure safety on the trail. Their distance from the western sections of the Santa Fe Trail made them a poor choice to defend the trail and its travelers.

Established on October 22, 1859, by Captain Henry W. Wessels of the Second U.S. Infantry, Camp Pawnee Fork was situated near water, timber and grass—three ingredients necessary for the survival of any military post

Infantry troops at Fort Larned. Fort Larned, the guardian of the Santa Fe Trail, was an integral Indian annuity distribution post after the Medicine Lodge Treaty was negotiated. *Courtesy Deb Bisel, author's collection.*

on the Kansas plains. After its initial hasty construction, the post would be moved three miles from its original location and relocated to the right bank of the Pawnee River, roughly eight miles above its confluence with the Arkansas River. Camp Pawnee Fork would be christened Camp Alert on February 1, 1860. On May 29, 1860, the post was rechristened Fort Larned in honor of Benjamin F. Larned, paymaster general of the army.[28] Established as a guardian of the Santa Fe Trail, the post was ideally situated to guard the wet and dry routes of the trail. As with any frontier post, the soldiers initially stationed there would be chiefly responsible for the construction of the post. Sod and adobe structures were quickly constructed, and soldiers lived more like prairie dogs than men. Writing in his diary, Captain Lambert Wolf noted on October 23, 1859, "plans are made for the horse and cattle stable, also for officer's and company quarters, all of which are to be built of sod, cut with spades by the members of our company. Our stable is to be 100 feet square...wall 12 feet high."[29] With time and an increase in troop strength at the post, locally quarried sandstone and backbreaking work would yield a new, permanent post that was the jewel of the West.

Initially, the post's duties centered on the protection of the mail and goods transported along the Santa Fe Trail. During the Civil War, ensuring the trail's security and halting Native American attacks were the prime objectives of soldiers stationed at Fort Larned. By 1865, the original sod and adobe dwellings were being replaced with sturdy, handsome sandstone structures. Born in Leipzig, Germany, Private Adolph Hunius served in the Union army during the Civil War. At the close of the war, he reenlisted and was stationed at Fort Larned. His diary—often times filled with humor—tells the story of Fort Larned from one soldier's perspective. In 1867, while at the post, Hunius was at times on work detail in the stone quarry. On June 12, he noted in his diary, "This morning to the Stonequarry. I on stone wagon. Gunning and Morris had to walk water rising. Dinner I cook. Very windy. I build a fire place to light a pipe."[30] Such was a soldier's life on the Kansas plains.

As white settlement in the area around the fort increased, competition for natural resources—land, timber, water and bison—increased. With settlers and Native Americans all clamoring for the same resources, tension on the plains increased. In the fall of 1864, relations deteriorated after Colonel John M. Chivington and his Colorado troops swept through Sand Creek attacking the Cheyenne encampment under the leadership of Black Kettle and White Antelope. From Fort Larned, Indian agent J.H. Leavenworth voiced his horror at the Chivington affair: "It is impossible for me to express to you the horror with which I view this transaction; it has destroyed the last vestige of confidence between the red and white man…what can be done? Nothing."[31] Along with Indian agent Edward W. Wynkoop, Leavenworth attempted to negotiate peace with the Kiowa, Comanche and Cheyenne. Fort Larned served as the base of operations for both men as they sought to prevent an all-out Indian war on the plains. In 1867, continued raiding and depredations on the part of the tribes made the situation untenable. General Winfield

Fort Larned National Historic Site as it appears today. The post is one the best examples of a Plains Indian Wars post. *Courtesy Michelle M. Martin.*

33

S. Hancock, the Union hero of Gettysburg, came to Fort Larned to quell the rising tide of Native American unrest.

General Hancock, whose success at Gettysburg was much lauded, underestimated the strength, resolve and resourcefulness of his new enemy. After an unsuccessful attempt at negotiating peaceful relations at the Cheyenne village along the Pawnee Fork on April 13, 1867, Hancock himself touched off a new round of conflict. In a show of force the next day, Hancock ordered the village burned and then dispatched General George Armstrong Custer to pursue the Cheyenne that had jumped the village the night before. Hancock's War had begun. By the fall of 1867, however, the various chiefs on the southern plains had been induced to gather for a great peace treaty council. Fort Larned would play a major role in the council and the fulfillment of its terms once negotiated.

On October 16, 1867, Captain Albert Barnitz of the Seventh U.S Cavalry wrote to his wife, Jennie, from a temporary camp on Medicine Loge Creek. Having just passed through Fort Larned, Barnitz witnessed the grandest gathering of Plains Indian leaders ever seen.

> *Well, there are a great many Indians here now—probably 7,000—in all, big and little and there are daily accessions to the number. A large band of Comanches came in yesterday and a band of Cheyennes last night. The big Council will come off on Saturday, and a general peace will doubtless be concluded,—which will last till spring, I suppose!—though the Commissioners seem to believe that it will be permanent. Possibly it will hold for a longer period that I imagine, and I do hope it will.*[32]

At the conclusion of the Medicine Lodge Treaty Council, the tribes agreed to give up land and the ability to roam in return for life on reservations and annual annuity payments in the form of money and goods provided by the Great White Father in Washington, D.C. Fort Larned would enter a new phase in its life and become a critical annuity distribution center pursuant to the terms of the Medicine Lodge Treaty. Like all treaties, however, the peace it was intended to foster would not last. Fort Larned continued to grow and troops moved to and fro in efforts to civilize, tame and conquer the West. Barnitz provides an intimate and eloquent look at life on the post in this time period. Writing to his darling Jennie, he talks of the mundane aspects of a soldier's life as well as providing vivid descriptions of the post: "Last night we had a dance on the parade, in front of Head Quarters, all the ladies from the Post were here. We had the 'flys' of a large hospital [tent] stretched on

Historical reenactors recreate the Medicine Lodge Peace Treaty proceedings every five years in Medicine Lodge, Kansas. The treaty council boasted the largest gathering of Southern Plains chiefs. *Courtesy Michelle M. Martin.*

the grass, and staked down for carpet. The music was very fine. The clog dancers of 'E' Troop in their brilliant costumes, gave an exhibition of their abilities that evening."[33] One can picture the officers and ladies dancing on the parade ground against the majesty of a Kansas night sky.

As with other posts of the nineteenth century, Fort Larned—despite its permanent sandstone structures and storied past—waned in importance to the mission of the army in the West. Described as a model of neatness and strict military discipline, the post that had been the guardian of the Santa Fe Trail was shuttered on October 3, 1878. The buildings of the post would be sold. The former post structures and land changed hands numerous times. In 1902, E.E. Frizzel bought the Fort Larned ranch and all the former post buildings. Over the years, the buildings were modified but still remained intact. In 1964, after much effort, Fort Larned became a unit of the National Park Service. Today visitors can tour the original post structures and walk in the footsteps of George Armstrong Custer, Winfield Scott Hancock, H.M. Stanley, Buffalo Bill Cody, Kit Carson, Edward Wynkoop and General Philip Sheridan.

FORT LEAVENWORTH: 1827–PRESENT

By the 1820s, America was headed west. The 1803 Louisiana Purchase had given the United States ownership of the rich interior of the North American continent. Reports from the Lewis and Clark expedition had excited curiosity and stirred thoughts of the opportunity to be had in that vast area. America was eager to trade with Mexico, and both countries pushed to relax trade restrictions that had forbidden it. The Mexican Revolution of 1821 did just that, and travel immediately sprang up from Missouri to Santa Fe. The only obstacle to a flourishing trade was the Native American population, which was not happy about the increased traffic across their lands. A treaty between the United States and the Osage Nation led to the establishment of the Santa Fe Trail in 1825. The Santa Fe Trail ushered in the era of everything that has been romanticized about the West. The forts, the wagon trains, the cowboys, the cattle drives, the cavalry charges and the Indian attacks—all of this resulted from the establishment of this historic trail. It roughly divided the area that would become the state of Kansas in half diagonally. Kansas has more miles of the trail than any other state.

As traffic along the trail increased, so did conflicts with Native Americans. Missouri had the greatest economic interest in maintaining safe commerce, so Missouri senator Thomas Hart Benton lobbied for protection and Secretary of War James Barbour authorized creation of a fort in late 1826. A location on the Arkansas River was considered but deemed impractical. It was decided that the best location would be the confluence of the Little Platte and Missouri Rivers, which would be near the beginning of the Santa Fe Trail at Westport, Missouri (present-day Kansas City). In the spring of 1827, Colonel Henry Leavenworth was sent from Jefferson Barracks, near St. Louis, to survey a site. The colonel quickly realized the area his superiors had chosen on paper was prone to flooding. Instead, he chose a spot twenty miles upriver and on the west side of the Missouri where high bluffs offered a suitable vantage point high above the water. A crude outpost sprang up almost overnight.

Cantonment Leavenworth was temporarily abandoned because of malaria in 1829. So many cases were reported that the inhabitants were ordered back to Jefferson Barracks. In November, the Sixth Infantry Regiment returned due to the numerous Indian attacks on the Santa Fe Trail. (It was thought that the winter would be safer since insects would not be an issue.) Soldiers from the post would accompany wagons to the Arkansas River, where a Mexican escort would meet them and travel on to Santa Fe.

A cannon perched atop the high ground overlooking the Missouri River at Fort Leavenworth. *Courtesy Michelle M. Martin.*

As if there were not enough problems with Native Americans, President Andrew Jackson added fuel to the fire with the Indian Removal Act in 1830. Eastern tribes were forced to leave their native lands for relocation in the West. This was another of those pivotal moments in American history where Kansas was at the center. Not only were soldiers from Leavenworth protecting whites from Indians, now they were also trying to keep the peace among the various tribes as well. Leavenworth was upgraded to a fort that same year. In 1834, the First Dragoons were stationed at Fort Leavenworth. This was the first cavalry unit in the U.S. Army.

The 1830s saw Texas declare its independence from Mexico, but friction remained. When the Mexican War broke out in the 1840s, the Army of the West, consisting of the First Dragoons and 860 Missouri Mounted Volunteers, was housed at Fort Leavenworth under the command of Colonel Stephen W. Kearny. (Kearny is the namesake of Fort Kearny, Nebraska.) "Kearny and his force of Dragoons and Missouri Volunteers departed Fort Leavenworth for New Mexico on June 26, 1846. A huge crowd of Army families and local well-wisher turned out to see the army off," wrote Lieutenant Colonel Phillip W. Childress. "It was an impressive force consisting of more than 100 wagons, 500 pack mules, 1,550 covered wagons, and a large herd of beef cattle. For the first time, soldiers at Fort Leavenworth took up the cudgel of war against a foreign foe."[34]

Childress went on to say, "Fort Leavenworth played an important role in the Mexican War, a war that propelled the fort into national prominence. It would never be an obscure frontier post."[35]

When the *Kansas-Nebraska Act* was signed in May 1854, the post served as the first capital of the new territory. During the years of unrest that followed, it would play a critical role in attempts to curb the violence that marked Bleeding Kansas and in defending the new state during the Civil War. Ironically, the last real threat during the war came from a man who had once served at this post—Confederate general Sterling Price. During the Mexican War, Price had served at Leavenworth in the U.S. Army and led an expedition into Mexico from the post. In 1864, he invaded Missouri in a last-ditch attempt to bring it into the Confederacy. One of his targets was the arsenal at Fort Leavenworth, but he was unsuccessful.

Following the Civil War, the fort was the staging ground for many engagements during the Plains Indian Wars. One of its more famous residents at the time was George Armstrong Custer. The Civil War's "boy general" had been court-martialed and suspended from service for a year. He and his wife, Libbie, were guests of General Phil Sheridan, who lived in the original sutler's home. Built in 1841, it is one of the most historic buildings on post.

When Kansas was admitted into the Union on January 29, 1861, the federal government failed to retain title to the fort and grounds. This was remedied by an act of the Kansas State Legislature on February 22, 1875, when the state ceded the area back to the United States.[36]

Another Civil War/Indian War personality to leave a lasting impact on this post was General William Tecumseh Sherman, long remembered for laying waste to the South. A less controversial and more far-reaching accomplishment was the establishment of the United States Infantry and Cavalry School in 1881. Sherman rightly observed the need for increased professionalism in army officers and chose Fort Leavenworth in part because of its distance from the nation's capital. Sherman's own experiences had left a bitter taste toward politicians, and he thought it best to remove the training of officers from their corrupting influence as much as practicable. The school he founded has evolved into the Command and General Staff College (CGSC).

In 1866, General Ulysses S. Grant directed the organizing of two cavalry regiments of black troops. Sherman made another notable contribution in the selection of Colonel Ben Grierson to organize the Tenth Cavalry at the post. The "Buffalo Soldiers" were memorialized by a subsequent black officer, General Colin Powell, who spearheaded the erection of a memorial statue.

The CGSC is only one of the reasons that Fort Leavenworth has been dubbed "the intellectual center of the army." The other entities of the Combined Arms Center are: School of Advanced Military Studies (SAMS), Command and General Staff School (CGSS), International Military Student Division (IMSD), School for Command Prep (SCP), School of Advanced Leadership and Tactics (SALT), Center for Army Leadership (CAL), Center for Army Profession and Ethic (CAPE), Combat Studies Institute (CSI),

Defense Language Institute (DLI), Military Review and Warrant Officer Career College (WOCC).

The Frontier Army Museum on post is dedicated to the conservation and acquisition of frontier artifacts. The Frontier Army Museum collection contains over seven thousand items, including weapons, uniforms, equipment and vehicles (including the carriage used by Abraham Lincoln when he visited the Kansas Territory in 1859 and a sleigh built for General Custer). Among the many premier artifacts are a dragoons' helmet (predates the cavalry soldier) from the 1820s that is one of only two known to exist and an 1832 general officer coat worn by General Henry Leavenworth.

For the same reason that Sherman thought an officers' school should be located at the post, officials considered the post ideal for locating a military prison. According to Major Gary Wade:

> *Military offenders had previously been confined in twenty-one different military stockades and eleven civilian penitentiaries in various states. Punishment varied from institution to institution and included flogging, use of ball and chain shackling, tattooing or branding, and solitary confinement. Early in the 1870s these punishments, except military confinement, were banned throughout the Army. But the 346 military prisoners in the civilian jails, however, suffered punishments prohibited by the Army. Major Thomas F. Barr (later brigadier general and known as the father of the U.S. Military Prison) advocated reform of the penal system. Based on Major Barr's proposal, a board of officers recommended that the U. S. Army adopt the British military prison system, whereby military prisoners would be consolidated at one location. An 1873 law authorized a military prison to be built at Rock Island Arsenal, Illinois. The Ordnance Department and Secretary of War William W. Belknap objected to the location, citing security reasons and stating that prisoner labor was ill-suited for munitions work. The secretary of war countered with the suggestion of Fort Leavenworth as the site for the military prison.[37]*

In May 1872, Congress authorized moving the prison to Fort Leavenworth and appropriated funds to that purpose. Operations at the United States Military Prison began in 1875. It was later changed to the United States Disciplinary Barracks (USDB) and is the only maximum-security correctional facility in the Department of Defense. It is the oldest penal institution in continuous operation in the Federal system.

FAMILY

For some folks, Fort Leavenworth is in the blood. Bill Buckner's life literally began there.

"I was born in 1926 at Fort Leavenworth while my father was stationed there [1924–1927]," said Bill. "I subsequently moved with the family to five army posts, attended three high schools and attended the United States Military Academy [West Point]."[38]

Buckner has long been a familiar name in military circles. Simon Bolivar Buckner was a Confederate general from Kentucky and governor of that state. His son, Simon Bolivar Jr., was Bill's dad, and was attending the Command and General Staff School in the 1920s. The moves that Bill described reflected his dad's advancement in the army. His dad was promoted to brigadier general in 1940, major general in 1941 and lieutenant general in 1943.

On June 18, 1945, Simon Bolivar Buckner Jr. was watching the last combat operations in the Battle of Okinawa when he was hit by shrapnel and killed instantly. He was the highest-ranking United States officer to die by enemy fire. He was buried on Okinawa and eventually removed to the family plot in Frankfurt, Kentucky. Buckner was posthumously promoted to general in 1954.

Bill graduated from West Point in 1948 and then headed to Fort Riley.

"I liked this post and my fellow lieutenants, despite having to crawl through scratchy sunflower fields during night compass training," he said. "From Fort Riley, I was transferred to Fort Knox, Kentucky, to take basic Armor [tank] training. My next three years of duty [1949–1952] were with the First Infantry Division in Germany. In Germany, we were always anxious about Russian combat intentions since the Russians would maneuver aggressively

General Simon Bolivar Buckner II while stationed in Alaska. The Buckner family's military roots run deep at Forts Atkinson, Leavenworth and Riley in Kansas. *Courtesy the Buckner Family.*

up to the East-West border from time to time—and stop. Back in the States for the next couple of years, I attended more schools [Marine Amphibious Warfare and advanced Armor training]. In 1954, fighting in Korea had ceased, I was still unmarried and I began to lose interest in continuing an army career. I had been in the service for six years, which was two more years than I was required. I resigned my commission in the Regular Army, married a Kansas City beauty and stayed in the reserves two more years while I pursued a business career."[39]

For Lieutenant General (retired) Don Holder, his memories of the storied post begin in childhood as well, but his interest never waned. He was committed to a military career.

"My father, Colonel Leonard D. Holder, attended the special weapons course at Fort Leavenworth before he went off to the Korean War," said Don. "That was in the fall and winter of 1951. We lived downtown for the short duration of that course."[40]

The elder Holder was back at Fort Leavenworth in 1956.

"Back then, the college held its classes in Gruber Hall," added Don, "the former riding hall that's now a fitness center. We lived in converted barracks, four families to a building, in West Normandy. [Hoge Hall now occupies most of that acreage.] Students worked hard back then. I remember that he spent a lot of time writing a study of Jackson's Valley Campaign for that course and that my mother—like most of the wives of students—put hours into layer-tinting maps for him. Student military maps were printed in black on white paper. Layer-tinting is hand inking of contour intervals to make the high and low ground stand out. The college bookstore sold India ink pots in sets that ranged in color from pale yellow to deep red."[41]

His father's career ended when he was killed in action in Vietnam in 1968. The time spent at Fort Leavenworth left Don not only with good memories but also with a real understanding of what it takes to make a professional soldier. In time, he would leave his own imprint on the post:

"I had five assignments at Leavenworth. I was in the CGSC class of 1977, returned to write doctrine at DTAC from 1980–82 and came back again for my Army War College fellowship at SAMS in academic year 1984–85. Subsequently, I was the director of SAMS from 1987 to 1989 and, finally, commandant of the college/commander of CAC from 1995–1997.

In March 2009, the CGSC hosted visitors from the British Army equivalent in Shrivenham, England. The British majors were impressed by the Liberty Memorial in Kansas City, the culmination of their American tour. *From left to right:* LTC Scott Porter (USA, Ret.), National WWI Museum and Liberty Memorial Board of Trustees; COL Robert E. Ulin (USA, Ret.), CEO, U.S. Army Command and General Staff Foundation; LTG John E. Miller (USA, Ret.), Vice Chairman, U.S. Army Command and General Staff Foundation; MG James Bashall, Deputy Commandant, Joint Services Command and Staff College (United Kingdom); and LTG Robert Arter (USA, Ret.), Chairman, U.S. Army Command and General Staff Foundation. MG Bashall eloquently commented on the mutual commitment of the United States and the U.K. to liberty, a word that is not used often enough, in his opinion. The CGSC has extensive collaboration with foreign countries. *Courtesy Deb Bisel.*

"The history of Fort Leavenworth fascinated me from the time I first lived at the post as a twelve-year-old. That interest in the frontier garrison and the migration trails that passed through it fed my curiosity about how Leavenworth's soldiers and schools influenced our history. By the time I returned there as a faculty member, I'd developed a strong appreciation for how much Leavenworth's schools and centers contributed to forming leaders for peace and war. My understanding of that led me to work hard later to foster greater regard for army schooling in the field and to improve the rigor and consistency of army training and doctrine."[42]

The retired general continues to be a frequent visitor at the post, teaching or consulting.

The old stone and brick "castle" was replaced by a new state-of-the-art, 515-bed facility in 2002. The USDB staff includes both civilian and service members of the Military Police Corps, Adjutant General Corps, Medical Corps, Medical Service Corps, Corps of Engineers, Chaplain Corps, Judge Advocate General Corps, the United States Marine Corps, the United States Air Force and the United States Navy.

The USDB maintains a small cemetery on post, with 298 graves of soldiers, including German POWs who were executed in 1945 after being convicted of murdering fellow prisoners. Vincent S. Green chronicled this incident in the 1995 book *Extreme Justice*. The prisoners' cemetery is not part of the national cemetery also located on the post. In the early years of Cantonment Leavenworth, tradition segregated officers and enlisted personnel. In 1858, remains from both those cemeteries were reinterred in what became one of the original fourteen national cemeteries either designated or established in 1862.

Numerous other small cemeteries throughout the West have had remains removed to Fort Leavenworth's national cemetery, including soldiers from Fort Craig, New Mexico, whose removal was necessitated by the building of the Atchison, Topeka and Santa Fe Railroad. The grave of Brigadier General Henry Leavenworth was relocated from Delhi, New York, in 1902. A large granite monument topped by an eagle marks his resting place. Notable interments include Captain Thomas Custer, who died with his brothers, nephew and brother-in-law at the Little Bighorn in 1876. Another of note is Lieutenant John Grattan, an inexperienced young officer who was killed near Fort Laramie, Wyoming, in 1854. He was attempting to arrest a Teton Sioux who had shot someone's cow when the tense situation escalated into the massacre of Grattan and thirty of his men.

In 2009, Command Sergeant Major Phillip Johndrow addressed the crowd assembled for Memorial Day:

> *Edmund Burke, a British statesman and orator, once said, "The only thing necessary for the triumph of evil is for good men to do nothing." Look around you. These gentle, rolling hills are dotted with the simple white headstones of men and women who were determined that good would overcome evil… and they put their lives on the line to ensure our way of life. To all of our veterans, we are deeply indebted.* [13]

Johndrow was correct in noting that it is the lives of these honored dead that are to be remembered. The contributions of the individuals who have served at the post are incalculable. Virtually every name that is now legend was educated at this fort. Dwight D. Eisenhower studied here, a year behind George Patton, whose notes he apparently used. They must have been pretty helpful because Ike graduated at the top of his class in 1926. In fact, when considering the leadership so vital to the Allied victory in World War II, the Command and General Staff College alums figure prominently: George C. Marshall, Douglas McArthur, Omar Bradley, Henry "Hap" Arnold, Simon Bolivar Buckner Jr., Walter Bedell Smith, Carl Spaatz and Mark Clark, to name just a few. The tradition continues as the nation, and the world, looks to Fort Leavenworth for leadership.

FORT MACKAY: 1845–1854

See Fort Atkinson.

FORT MANN: 1847–1850

Unlike our modern highway system that features clean, well-lit and handsomely appointed rest stops for weary travelers, the overland trails and routes of the nineteenth century were lacking in luxury. Creaking, lumbering and lurching across the prairies of Kansas, wagon trains were apt to break down. Soldiers on patrol far from home base needed places for respite. Travelers along the Santa Fe Trail could see the remains of animals, wagons and unfortunate humans that had not fared well in their travels. In a letter to his superior officers, Captain William M.D. McKissack, an assistant quartermaster, stressed the need for a post in western Kansas: "In crossing the plains there is no means of securing Wagons that become unserviceable for want of repairs; generally the bands, tires, spokes, etc. become loose on account of the dryness of the atmosphere and having no means of repairs; in such cases the Wagons are abandoned…owing to the great number of Wagons abandoned on the plains I make arrangement to erect Wheelright, Smith & Store houses near the crossing of the Arkansas."[11]

Needing a location that was exactly halfway between Fort Leavenworth and Santa Fe, Fort Mann was established in April 1847. Situated on the north bank of the Arkansas River and twenty-five miles below the Cimarron

A simple marker denotes the locations of Fort Mann, Mackay and Atkinson just west of present-day Dodge City, Kansas. The posts were important in guarding the Santa Fe Trail. *Courtesy Michelle M. Martin.*

Crossing of the Santa Fe Trail, Fort Mann served the needs of the United States Army and travelers until the erection of Fort Atkinson in 1850. Built by teamsters under the direction of Daniel P. Mann, for whom the post was named, Fort Mann was never staffed with professional soldiers, but rather with teamsters. The frontier post consisted of four log structures with a stockade and was considered highly defensible. With one six-pound artillery piece, forty rounds of grapeshot, forty cannon cartridges and six rifles and muskets, the men at Fort Mann were armed to defend themselves against the attacks from Plains tribes that became regular occurrences. The post was occasionally occupied by regular army troops and saw a great deal of tenuous engagement against various Plains tribes.

By 1848, Fort Mann had proved to be too isolated and far from other garrisons for it to be useful to the military. Maintaining the morale of those stationed at the post was difficult at best, and it was abandoned. Today, nothing remains of Fort Mann. A signpost just three miles west of present-day Dodge City commemorates the post and its role in attempting to supply wagon trains and troops and keep the peace on the western edge of civilization in Kansas.

FORT MONTGOMERY: 1855–1861

His lean face and long beard distinguished him from many of his contemporaries. A look of righteous fire burned in his eyes. James Montgomery came to Kansas after the passage of the Kansas Nebraska Act and settled on a land claim in Mound City, Kansas Territory. Quickly Montgomery was recognized as one of the most influential and zealous leaders of the Kansas free-state movement. Working alongside James Henry Lane and John Brown, Montgomery sought to ensure that Kansas would enter the Union as a free state. He, along with his supporters, was willing to lay down his life for the cause, and Montgomery's home became a veritable armed fortress. For Montgomery, life was precarious. His first home had been burned and the safety of his family became his paramount concern. "There was scarcely a week without some ugly crime to bring distress and indignation to the families of all the free-state settlers. Montgomery was frequently shot at and his numerous escapes made him thoughtful to secure better protection for himself and family."[15]

Built of oak and walnut, Montgomery's two-story cabin became the center of free-state activities in Linn County, Kansas Territory. Perched atop a hill, giving him the ability to see all coming toward his fortress, Montgomery's home/fort boasted portholes on the second floor that served as rifle ports. With thick walls, rifle ports and a commanding position atop a hill, his home/fort survived not only the turmoil of Bleeding Kansas but the Civil War as well. From this fort, he planned and carried out his military operations in the territory, including two raids on the nearby community and former military post of Fort Scott.

On December 9, 1858, Montgomery led his men to the predominantly proslavery community of Fort Scott to free one of their comrades, Benjamin Rice. Rice had been jailed in one of the former officer's quarters that had been converted to a hotel after the fort's closure in 1853. During the nighttime raid, Montgomery and his men encountered resistance from citizens, including former U.S. deputy marshal John Little. After the smoke cleared, Little lay dead in his family's home, the former post headquarters. In the aftermath of Montgomery's raid, Sene Campbell, the fiancée of the murdered Little, wrote to Montgomery. Her letter, printed in the *Fort Scott Democrat*, laid her anguish bare for all to read. One wonders how Montgomery felt when he read her words and whether he was seated before his fireplace in his warm home/fort, surrounded by family, as Campbell called him "a minister of the Devil and a very superior one too."[16] The need for his fort/home was well

Historical interpreter Greg Higginbotham portrays James Montgomery raiding Fort Scott during the Kansas territorial period. *Courtesy Michelle M. Martin.*

founded, for the diminutive Campbell chided Montgomery and made her intentions toward him quite evident in stating, "But remember this. I am a girl, but I can fire a pistol. And if ever the time comes, I will send some of you to the place where there is weeping and gnashing of teeth."[47]

After his death in 1871, the home of James Montgomery fell into ruin. Today, the citizens of Mound City, Kansas, are proud of their reconstruction of Montgomery's home that stands in their historical park.

FORT RILEY (CAMP CENTER): 1853–PRESENT

As the West became a focal point for travelers and settlers in the mid-1850s, the troublesome task of protecting these groups fell to the men of the United States Army. With the California, Santa Fe and Oregon Trails all traversing Kansas soil and numerous waterways serving as conduits of transportation, Fort Riley was ideally situated. Located at the confluence of the Smoky

Hill and Republican Rivers, Fort Riley was founded in 1853, just one year prior to the passage of the Kansas Nebraska Act, and continues to serve our nation today.

By 1851, the nature of the Permanent Indian Frontier was changing. With increased white settlement in the west, Colonel Thomas Fauntleroy, the commander of the First U.S. Dragoon regiment, declared that Forts Leavenworth, Scott and Atkinson in Kansas should be closed and new posts constructed farther west. With support from Secretary of War Charles Conrad, Fauntleroy was granted one of his wishes—a new location for a fort in Kansas would be scouted and a post constructed at the junction of the Republican and Smoky Hill Rivers in north central Kansas territory.[18] The role of this potential new fort was twofold: to provide the safety needed by those traversing through Kansas along the trails westward and to begin the pacification of the Plains tribes.

During the spring of 1853, construction began on the new post. Captain Edmund A. Ogden, Quartermaster Division, oversaw the construction process and intended on using the soldiers assigned to the post as his main source of labor. This practice was common in the nineteenth-century army, and other posts in Kansas, including Forts Leavenworth, Scott, Larned

The historic cemetery at Fort Riley. Along with service personnel, civilians that perished on post are interred here. *Courtesy Michelle M. Martin.*

and Wallace, would be constructed similarly. With troops of the Sixth U.S. Infantry and civilian contractors working together, the post began to take shape ever so slowly. Rock was quarried locally for the construction of post buildings. Beset with labor shortfalls, construction moved at a tedious pace. The use of active troops as the main labor force to construct the post slowed building progress as their regular military duties took them away from construction detail.

By July 4, 1853, a visitor to the region described the post as being in a "half finished state, already making a fine appearance from a distance."[19] Troops living in tents assuredly differed in their opinion and longed to sleep in more comfortable accommodations. Shortly after their Independence Day celebrations in 1853, the post received word that Major General Bennett C. Riley had died. For his varied service to the nation, General Order No. 17 named this newly developing post Fort Riley.

As Kansas Territory became a breeding ground for hostility over the issue of slavery, Fort Riley would become a guardian of the trail, protector of settlers and agent of stability. For Fort Riley, however, stability was difficult in these early days. Continual labor and supply shortages hampered the progress needed to finish construction. In 1855, a new and more deadly foe wrought havoc on the construction of Fort Riley and threatened to stop its progress permanently—cholera. In a one-week period in July 1855, over 150 individuals lost their lives to the disease, including Major Ogden. Petrified workers fled the camp. Chaos ensued. By September 1855, however, work was moving forward, and in December of that year, Colonel Phillip St. George Cooke arrived and assumed command of Fort Riley. His leadership would set the post on a firm foundation. During this tumultuous period, young Lieutenant James Ewell Brown Stuart of Virginia arrived at Fort Riley with the First U.S. Cavalry. His time at Fort Riley during Bleeding Kansas would serve as a training ground for the larger conflict that erupted in 1860 with the election of Abraham Lincoln to the presidency and the secession of the Southern states from the Union. While in Kansas, Stuart would hunt and enjoy "a fine roast of buffalo" and experience the natural wonder of a "bird serenade at night" on the Kansas prairie, all while searching for Native Americans to subdue.[50]

With the opening salvos of the American Civil War, Fort Riley served as an outpost that protected the trails, settlers and critical commerce. The post would also send troops to defend Kansas's borders against the threat of attack from pro-Confederate partisans from Missouri. Fort Riley would serve as a training ground for Kansas units before they were shipped to the theater of war as well. With Kansas units and troops from other states shipped in and out of Fort Riley, the post was abuzz with activity during the war. The post and its strategic location made it perfect for training and mustering troops into Union service. In 1864, the addition of a new military district (District of the Upper Arkansas) required a centrally located post for its headquarters, and Fort Riley was perfect.[51] This ensured increased civilian employment on post, making it a vital part of the growing community of Junction City. As the war drew to a close, however, the future of Fort Riley hung in the balance. Another war—with an enemy that did not fight like regular army—would ensure the post's future.

While some Kansas forts faded into the pages of history after the close of the Civil War, Fort Riley was on the rise. To secure peace on the plains and allow for the settlement of the West to continue unabated, Fort Riley housed numerous units charged with the task of taming the West. In 1866, several new cavalry units would make Fort Riley their home: the Seventh, Eighth and Tenth U.S. Cavalry regiments arrived in Kansas ready to serve and protect in the West. The Seventh, of course, became famous for its flamboyant leader, General George Armstrong Custer. Fort Riley became the headquarters of the Seventh as it traversed the plains and points farther west

A statue honoring members of the famed Buffalo Soldiers, African American cavalry troops, at Fort Riley. The post was their headquarters. *Courtesy Michelle M. Martin.*

The famed Buffalo Soldiers at Fort Riley during World War I. *Courtesy United States Army.*

on campaigns against Native Americans. The Tenth would become known as the Buffalo Soldiers—African American troopers that fought tenaciously on the plains. They, too, would make Fort Riley their headquarters with the arrival of their commander, Colonel Benjamin Grierson, in 1867. As the Plains Indian wars drew to a close once again, Fort Riley would find itself searching for purpose on the plains.

After serving Kansas since its inception, by the 1880s Fort Riley was in need of repair. The construction of new barracks and repair of existing structures commenced. In 1882, the cavalry and artillery began operation, and Fort Riley assured its continuation into the new century. As 1890 dawned, only Forts Leavenworth and Riley stood guard, lonely sentinels on the Kansas prairie. Fort Riley's locale made it a natural choice for the army to establish schools of instruction—its large landholdings combined with its access to the rail system helped the military balance its mission with the best use of manpower and resources. Ironically, as the twentieth century drew closer, a witness to the past died at Fort Riley. Comanche, the faithful mount of Captain Myles Keogh of the Seventh Cavalry and a survivor from Custer's defeat at the famed Last Stand Hill, died at the post. With the passing of Comanche, Fort Riley, it seemed, was moving toward a brave new future.

In real estate, location is everything, and the same holds true for the military. Fort Riley—with its access to manpower, land, rail lines and communication and its ability to operate at a lower cost—was an ideal permanent post. In the twentieth century, the post served to train troops for combat. With American entry into World War I, the parade and drilling grounds at Fort Riley were filled with men learning the bayonet drill and the fine art of soldiering.

As the art of war advanced, new and frightening weapons would supplant the horse-mounted soldier. Fort Riley adapted its training mission and changed with the times. By 1937, the Seventh Cavalry Brigade (mechanized) made the post its home. During the Second World War, the post would house a prisoner of war camp and send troops into combat. The post–World War II era saw numerous mission changes for the army. With the threat of nuclear war and a new enemy—the Soviet Union—Fort Riley formed the Aggressor School. The school trained soldiers in doctrine and tactics. Participants used role-playing to simulate combat against a new enemy. Speaking a specially created language and wearing unique uniforms, these troops, once trained, traveled to other bases and assisted in training other units.[52] The addition of the Army General School (AGS) in 1950 ensured that a new generation of

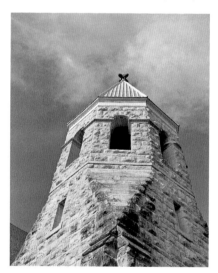

The tower of the historic chapel located at Fort Riley. Fort Riley saw action from the tumultuous days of Bleeding Kansas and today is the home of the venerated First Infantry Division, the Big Red One. *Courtesy Michelle M. Martin.*

officers received training. With schools of intelligence, counterintelligence and tactics, Fort Riley has trained generations of military professionals to serve, protect and defend America.

Throughout its storied history, numerous units have called Fort Riley their home. From the famed and ill-fated Seventh Cavalry of the nineteenth century and the venerated Buffalo Soldiers of the Tenth to the infantry men of the Big Red One, Fort Riley has been at the epicenter of American military life and culture. Today, troops from the post serve both at home and abroad. Visitors to the post can literally walk in history's long shadow and see the remnants of the nineteenth-century post and today's side by side.

FORT SAUNDERS: 1854–1861

Fort Saunders was one of many proslavery camps that were given the moniker "fort" during the tumultuous years of Bleeding Kansas. Situated twelve miles southwest of Lawrence, in Douglas County, Fort Saunders presented a continual threat to free-state settlers in the region. A base from which men could resupply and find shelter, Fort Saunders was the scene of cold-blooded murder.

D.S. Hoyt of Lawrence rode to Fort Saunders on August 12, 1856, to enlist the aid of Colonel B.F. Treadwell, the commander of the post. Raiders using Fort Saunders as their operational base had been pillaging around the nearby community of Franklin. Upon leaving the camp, Hoyt was murdered. This act touched off a new wave of violence that resulted in a force of men on horseback and foot leaving Lawrence to make their way to Franklin to put an end to proslavery raiding and thievery once and for all. Led by James Henry Lane on August 15, 1856, a group of free-state men made their way to Fort Saunders. Their advance was noticed by the men inside Fort Saunders, and Colonel Henry Treadwell, Saunders' commander, and his men fled rather than fight. Fort Saunders burned brightly as Treadwell's men made their way into the countryside. Like many of the temporary forts and bases from the territorial period, nothing remains of Fort Saunders today.

FORT SCOTT: 1842–1854 AND 1861–1865

As members of the Cherokee and Creek Nations made their way west to Indian Territory, they discovered their newly assigned homeland was contested ground. Tribes indigenous to Kansas, Nebraska and Oklahoma watched as newcomers began to settle lands they had been traversing for generations. To add to the turmoil among the tribes, many poor white settlers in search of free land saw Western lands as a panacea for their woes. Ensuring peace and stability on the frontier became one of the main objectives of the United States Army in the West. By establishing a line of permanent military forts along the "permanent Indian frontier" line in the West, white civilization could be kept at bay and peace kept among the tribes as they learned to accept their new living conditions. First envisioned by Secretary of War Joel Poinsett in 1837, the permanent Indian frontier line would be a series of fortifications that stretched from Fort Snelling

Mexican War reenactors fire a twelve-pound mountain howitzer at Fort Scott National Historic Site. *Courtesy Michelle M. Martin.*

in Minnesota to Fort Jessup in Louisiana.[53] The land that would become Kansas would play a critical role in this line of defenses.

One of the key factors supporting the foundation of Fort Scott was the need to keep the peace among the Osage and the newly arrived tribes in the region, including the Cherokee. A post in the neutral lands in what would become Kansas would prove invaluable in this mission. If constructed close to the Missouri border, this would also aid in keeping white Missourians on their side of the boundary and slowing encroachment into neutral and tribal lands. In addition, supplies and manpower demanded another post on the frontier.

Fort Leavenworth was the first military outpost established in what would become Kansas. Following the line south into present-day Oklahoma, Fort Gibson was a critical outpost. The distance from Fort Leavenworth to Fort Gibson made the transport of supplies and men difficult. Establishing another post at a halfway point would be invaluable to the army's mission of keeping the peace on the frontier. With the decommissioning of Fort Wayne in Cherokee lands, the need for Fort Scott was even more pressing.

In March 1842, Captain Benjamin Moore and Dr. J.R. Motte set out to select a site for a new frontier post. Water, timber and grass were three major requirements for any military post on the frontier. Finding all three along the banks of the Marmaton River in eastern Kansas, a location on the high prairie was selected and Fort Scott was born.

Named for General Winfield Scott, a military tactician, the post would serve as the needed half-way point between Forts Leavenworth and Gibson and go on to play a vital role in supplying the Army of the Southwest, sending troops to the Mexican War, and later became the nucleus of the town of Fort Scott, Kansas. During the Civil War, the post would be the scene of two historic firsts: the first induction of African American Union army regiments into service (First and Second Kansas Colored Infantry) and the creation of three regiments of Native American Home Guard (First, Second and Third Regiments Indian Home Guard). The fort would be the scene of triumph and tragedy from 1842 to 1854 and again from 1861 to 1865.

As Heiro T. Wilson made his way on post in 1843, he surely must have been taken aback as he gazed at the sight that greeted him. Wilson "found the officers and soldiers in one story log houses, the chinks filled with common mud, no floors."[51] For soldiers garrisoned at Fort Scott in its early days, the task of building the post fell squarely on their shoulders. Needing leadership and guidance to manage men and materials, the army would send Captain Thomas Swords, First Dragoons, to oversee the task of building the post.

Army mules and wagons rumble across the parade ground at Fort Scott National Historic Site. The post is restored to its 1840s appearance. *Courtesy Michelle M. Martin.*

His letters to his friend Lieutenant Abraham R. Johnston chronicle his monumental task. "We are going to make it the crack post of the frontier," he wrote to Johnston.[55] Swords's vision for Fort Scott was grand. By the time construction was complete, the post structures (composed of stone and wood obtained locally) included four officer's quarters, a post command, three barracks, stables, a hospital, guardhouse, quartermaster supply, bakery, powder magazine, well canopy and mess hall. Swords lived up to his mission to build an exemplary post on the frontier.

Despite sending troops to guard the trails, ensure Native American peace on the frontier and participate in the conflict with Mexico, Fort Scott's most important days came after its closure in 1853. With the passage of the Kansas Nebraska Act in 1854, settlers flocked to Kansas Territory. Due to its proximity to the Missouri border, Fort Scott would be visited by violence and bloodshed from 1854 to 1861. With the entry of Kansas into the Union as a free state on January 29, 1861, the former post and growing community of Fort Scott would play an integral role in defending Kansas from Confederate invasion from Missouri.

Reoccupying many of the former post buildings and renting any available space in the community, soldiers once again walked the streets of Fort Scott during wartime. Units at Fort Scott would keep peace along the border, fight the Confederate enemy in the Indian Territory (Oklahoma), engage in battle in Missouri and Arkansas and defend Fort Scott from threatened invasion. More troops were stationed at Fort Scott during the Civil War than had ever been during her official "active" period. To assist in the defense of Fort Scott, a series of lunettes (military fortifications with two protective faces, two parallel flanks and a total of four sides) were constructed beginning in 1862. The objective of Forts Blair, Henning, Insley and Lincoln was simple—protect the roads leading in and out of Fort Scott allowing trade, commerce and military men and supplies unfettered travel through Bourbon County.

The town was a veritable fortress—a fortress that Confederate general Sterling Price attempted to occupy twice. In 1864, Price launched an ill-fated raid attempting to occupy St. Louis and Jefferson City for the Confederacy. When he did not take either city, he headed west toward Kansas City and fought a series of running battles, rear guard actions that culminated in the Battles of Westport and Mine Creek in October 1864. Fearing Price's army of nearly twelve thousand, commanders at Fort Scott issued orders to evacuate the city and, if needed, burn all military stores and structures lest they fall into Price's hands. A valiant Union charge on the rolling Kansas prairie near Mine Creek in Linn County repulsed Price's men. The aftermath was horrid, and Fort Scott, saved from annihilation, tendered aid to the sick and wounded of the Battle of Mine Creek. Emma Caroline Morley, a young woman visiting family in Fort Scott, wrote in her diary:

Emma Caroline Morley lived in Fort Scott, Kansas, during Confederate general Sterling Price's 1864 raid. *Courtesy Matt Matthews, Kip Lindberg*

[T]*o me the sight was sad and sickening. Dead horses scattered all around, and cattle that had been hurriedly killed by the rebels, who not having time to cook all, had cut large pieces from them, leaving the unsightly remains to bleach on the prairie...Most of the dead we supposed had been buried, but saw two rebels left on the lonely prairie, with the scorching sun shining down upon their blackened up-turned faces. One was dressed neat and comfortable, fine features. Must have been a fine looking man. Someone doubtless loving and praying for him even now little thinking of his lying on this lonely prairie, the cold ground for a resting place, with nothing to protect him from drenching rain or scorching sun.*[56]

The sights of war hardened any delicate sensibilities Emma possessed. She, along with other townswomen, nursed the sick and wounded in the old post hospital and the tents set up on the parade ground to handle the influx of war refugees, contraband and soldiers. She continued her description in her diary:

Sad and sickening was the sight (in going through the wards) of our brave soldiers suffering from frightful and ghastly wounds. All seemed patient and cheerful, but death will soon relieve many of their sufferings. Went through the rebel ward and talked with many of these wounded prisoners. Those I talked with expressed themselves as tired of the war, tired of fighting. One young man whom I spoke to on coming out, seemed very sad, tears were in his eyes. Poor fellow, he is badly wounded. His cot, I fear, will soon be empty...This cruel war, when will it end? The broken hearts it has already caused...alas! How many more it will cause, the future alone can tell...[57]

Fort Scott housed refugees, contraband and sick and wounded soldiers during the American Civil War. Historical reenactors portray refugees at Fort Scott National Historic Site. *Courtesy Michelle M. Martin.*

Fort Scott had risen to the occasion and answered the call to service that was required to preserve the Union. After the conclusion of the war, Fort Scott was once again a civilian community.

By the late 1860s, railroads were being built at a breakneck pace in Kansas. After selling their lands in Southeast Kansas to the government, the Cherokee threat in the region was neutralized. However, many whites had settled (albeit illegally) in the neutral lands. Only the outbreak of the war and other more pressing matters kept the military from evicting them. Now, rail survey crews were a prime target for attack from angry settlers. Fort Scott would be headquarters for the Post of Southeast Kansas from 1869 to 1873 to protect railroad crews and their work. This ended Fort Scott's long and varied military history. The buildings of the old post became homes and shops as the town grew. In 1978, Congress and the Department of the Interior established Fort Scott National Historic Site to preserve the rich history of the former post. Today, the historic site features original refurbished and reconstructed structures that take the visitor back in time.

Fort Titus: 1855–1856

With the capture and destruction of proslavery Camp Saunders in Douglas County, Kansas Territory, complete, free-state forces were on the move in August 1856. Proslavery settlers in and around Lecompton lived nervously, waiting and watching for the next armed conflict with their free-state foes. Emboldened by their victory, free-state forces turned their sights toward Fort Titus on August 15, 1856. At dawn's first light, fifty free-state fighters divided into two parties and surrounded the small fortified structure. Built by Colonel Henry T. Titus in 1855, Fort Titus was nothing more than a reinforced log home located two miles south of Lecompton, the proslavery capital of the territory. Used as a rendezvous point and stronghold for proslavery forces in the area, Fort Titus would be a great prize for the free-state forces to capture.

In the battle that unfolded, an eyewitness recounted that "early in the morning a party of Free-State cavalry made a charge upon some tents

A replica of Fort Titus stands in a park near Lane University in historic Lecompton, Kansas. Lecompton was the proslavery capital of Kansas Territory. *Courtesy Michelle M. Martin.*

near the cabin, the inmates of which ran for the cabin, and were followed by the horsemen, who went too near the cabin, when they were fired upon by those inside…The cannon was then brought up, and Cpt. Bickerton coolly brought his piece to bear upon it. Seven balls had been fired into it, when Col. Titus showed the white flag, and surrendered."[58] The gunner is reported to have uttered the phrase, "This is the second edition of the Herald of Freedom" in reference to the antislavery newspaper in Lawrence that had been destroyed by proslavery forces earlier. Thus the Battle of Fort Titus ended. Free-state forces captured four hundred muskets, knives, horses, wagons, farm equipment, household provisions and $10,000 in gold and bank drafts. Titus's fort was burned to the ground, never to be rebuilt again.

Lieutenant James Ewell Brown (J.E.B.) Stuart spent a portion of his pre–Civil War military career in territorial Kansas. His experiences at Fort Riley, Lecompton and Fort Wise would ready him for combat in the Civil War. J.E.B. Stuart, WICR 31423 in the Collection of Wilson's Creek National Battlefield. *Image courtesy of the National Park Service.*

Today, a replica of Fort Titus greets visitors as they make their way into the historic town of Lecompton. Every two years, living history reenactors recreate the battle for eager crowds that embrace the turbulent history of the town and it's storied past.

FORT WISE (FORT LYON): 1859–1862

The high prairie of western Kansas Territory was a world away from the swirling politics of Washington, D.C. Underneath the unfinished dome of the capitol building itself, the bluffing and blustering of political enemies had come to a head. Following the election of Abraham Lincoln in November 1860, the nation could no longer forestall its differences. South Carolina seceded from the Union, and other Southern states began to fall in line as well. In the Kansas Territory, a new fort in the line of protection for the Santa Fe Trail had been named in honor of Virginia's governor, Henry Wise, in hopes this would encourage the border state to remain in the Union. The fort on the Arkansas River was actually built by William Bent in 1853 (Bent's New Fort) and leased to the federal government in 1859. Briefly called Fort Fauntleroy, it was soon renamed Fort Wise.[59] Wintering on the plains was never easy, but Lieutenant James Ewell Brown Stuart was concerned more about the fate of the country and his family when he wrote to his older brother, Alex, who was back in Virginia:

Fort Wise, Kansas Territory
January 18, 1861

My Dear Brother

Events are transpiring so rapidly that furnish no little hope of perpetuating the Union, that I feel it incumbent upon me to tell you my course of conduct in such a contingency. Of course I go with Virginia whether she be alone or otherwise but I am sure that a large military force will be required for a time by the state and I am anxious to secure from Wythe a legion of Cavalry (200 men) myself as commander, or a battery of light Artillery 100 men or less with Governor Letcher as governor and you on the spot I think I ought to be able to get such a command. I could soon drill and discipline such a corps—and with such material I would feel sure of making "Stuart's legion" as formidable as "Lee's legion" of the Revolution. As I can not be on the spot till after Virginia's leaving the Union I will have to get you to get up or assist in getting up such a corps…Of course till the time comes to act I shall expect you to keep this "subrosa."

I am very much annoyed at being separated from my family in these critical times. I fear I can't get down till April to Fort Riley…I made a temperance speech here Christmas which gained me great éclat among the officers and soldiers—there are one fourth of the command sons of Temperance; they had a grand procession and ovation I had only a few days notice and spoke twenty minutes.

In these times of trial our only hope is a firm reliance on Divine Providence by whose guidance all things will be bright.

Flora and the children were at Riley and well at last accounts…

Yours affectionately,
J.E.B. Stuart[60]

Stuart's situation would be mirrored in the fate of the fort itself. When Virginia did, in fact, secede from the Union, it was a little embarrassing to have a Union fort named for the governor of a Confederate state. On June 25, 1862, the name was changed to honor General Nathaniel Lyon, killed at Wilson's Creek, Missouri, in 1861.

Fort Wise was in the Kansas Territory until January 29, 1861, when Kansas was admitted to the Union, just days after Stuart's letter. The western border of the Kansas Territory had extended to the Continental Divide, but the state boundary line excluded the mountains and high plains that included Fort Lyon. Thereafter, it was a part of Colorado Territory and, in 1876, the state of Colorado.

In 1866, Arkansas River flooding undermined the fort, and the next year, a new Fort Lyon was built on the north bank of the Arkansas. A cemetery was established at the post in 1887, but the fort was abandoned in 1897 and the remains buried in the cemetery were transferred to Fort McPherson National Cemetery in Nebraska. In 1906, the fort buildings were converted into a sanatorium to treat soldiers and prisoners of war with tuberculosis, and burials began again in 1907. That cemetery is in the national cemetery system.

TRADING POST: 1854–1861

The location was excellent. A scant four miles from the Missouri border along the Marais des Cygnes River, Michel Giraud and Philip Choteau had found the perfect setting for a trading post. A member of the famed Choteau fur trading family, Philip saw the potential of trade with the Osage Indians in the region and in 1838 founded a lucrative business in the Indian lands. Early in the history of the region, only military personnel, missionaries and trading post operators could inhabit the unassigned Indian lands. With the establishment of Fort Leavenworth to the north and Fort Scott to the south, the post would take on a new role and assist in providing supply and protection for army troops on the move in eastern Kansas. In 1848, Choteau sold the post to Phillip Avery, and it continued to operate.

By 1854 when Kansas was opened for legal settlement by whites, the Choteau Trading Post had been operating for over a decade and would become a focal point for settlement. Trading with the Native American tribes and newly arriving settlers made the business even more profitable.

Settlers began to congregate in the area around the post and put down roots. One particularly fine spring day in May 1858 would forever change the lives of those that called the Trading Post area their home and inspire American poet John Greenleaf Whittier to take up his pen.

The morning of May 19, 1858, began like any other for the Stillwell family. William Stillwell had moved his family from Iowa to Kansas Territory and settled near Trading Post. With his young wife, Mary Jane, and their two children, the Stillwells were living the American dream. Owning a small plot of land, homesteading and running a mercantile business provided them with all they needed. As the sun shone brightly and the birds sang sweetly, William loaded his wagon and bade his family farewell as he headed to Kansas City to procure goods. Little did he know this was the last time he would look back and see his family again.

Across the Missouri line, Charles Hamelton waited for the right opportunity to strike at free-state settlers that had been the bane of every proslavery Missourian's existence. With a band of men, they saddled up and headed toward the Kansas line. Trading Post and the local inhabitants would not be safe this fateful day. Heavily armed and looking to settle past grudges, Hamelton and his posse began rounding up local men one by one. William Stillwell was met as he made his way north from Trading Post. By the time the riders completed their task, eleven men were marched into a deep ravine. Hamelton ordered the terrorized prisoners to lie facedown in the ravine. With horses and men atop the ravine, the men below realized they were about to die. As the shots rang out, men lay dead and dying while others gravely wounded clung to life.[61] The bravery of Sarah Read, the wife of one of the victims, saved those that hovered between life and death. In the end, five men lay dead, among them William A. Stillwell.

Historical interpreter Michelle Martin portrays Mary Jane Stillwell, one of the widows of the Marais des Cygnes Massacre. Stillwell lived near Trading Post before her husband's murder. *Courtesy Howard Duncan.*

Traveling through the contested territory in 1858, newspaperman W.P. Tomlinson heard the shocking news of death along the Marais des Cygnes.

Traveling to Mound City to witness the aftermath for himself, Tomlinson recorded the tragic impact of Bleeding Kansas for one woman and her children:

> *She wore a thick veil which prevented her features from being seen, but the convulsive sobs that momentarily shook her frame, gave unmistakable tokens of the most insupportable anguish of her mind under her sudden and terrible bereavement. The children too as though sensible of a father's loss, or from sympathy with their mother's uncontrollable sorrow, sobbed as though their little hearts would break. She would hysterically murmur: "O if he had only left some message for me, it would not have been quite so dreadful! O if he had only left some message of his wishes." The rude coffin was taken from the wagon and borne by stout men to its final place of rest. It was carefully lowered to the bottom of the grave, and the bearers stepped back to allow the grief stricken mourner a last lingering look at the receptacle containing all that was most cherished in life. She swept aside the thick folds of her veil, revealing as she gazed down into the narrow depths where her fondest hopes were soon to be buried forever, a face whose transcendent loveliness was visible.*[62]

Today, the trading post that became a territorial community has faded with the passage of time. A small pioneer cemetery marks the final resting place of four of the five Marais des Cygnes victims, and a local museum preserves the stories of Trading Post.

Chapter 2

AD ASTRA PER ASPERA: THE CIVIL WAR

1861–1865

*Kate, Kitty and I sitting around a fire about nine o'clock p.m. heard the band
playing and great cheering in camp. Was sure they had received good news
and was beginning to feel quite hopeful, when John came in to tell us the
rebels were at Mound City (about twenty miles from here) and our forces still
in the rear. My heart sank within me. I thought this place doomed, and could
see no chance of escape.*[63]
—*diary of Emma Caroline Morley, Sunday, October 23, 1864.*

Kansas was bathed in blood and fire from 1854 to 1861. With its entry
into the Union as a free state on January 29, 1861, the territorial fight
over the expansion of slavery into Kansas would become legitimized under
the banner of war. Kansans that had survived the tumult of the territorial
period hoped that peace would rule the land. Sadly, Kansas and its citizens
would be tested by war. When Abraham Lincoln called for seventy-five
thousand volunteers to squelch the rebellion in the cotton states, men from
Kansas answered the call like no other state in the Union. With its existing
military posts guarding the border and ensuring that Confederate armies in
Missouri, Arkansas and Texas would not encroach its boundaries, Kansas
needed new posts to train soldiers and press them into service for the war
effort and aid in the protection of the home front.

The colors of the First Kansas Colored Infantry. Kansas was the first state in the Union to actively recruit, train and field African American soldiers in the American Civil War. *Courtesy Deb Bisel.*

CAMP EWING: 1864–1865

From the time of its inception, Lawrence boasted a topographical advantage in battle—Mount Oread. As successive waves of proslavery partisans attacked the free-state stronghold, Mount Oread featured prominently in the defense of Lawrence. Home to Fort Lane in the territorial period, Mount Oread would play a significant role in the defense of Lawrence, but it would take a day of fire and death to make Mount Oread important once again.

Riding like demons from Missouri into Kansas, William Clark Quantrill and his forces had a score to settle with the "jayhawkers" of Lawrence, Kansas. Seeking revenge and retribution for wrongs committed in Missouri and the recent deaths of pro-Confederate women in the infamous Kansas City jail collapse, Quantrill's men wrought fire, death and destruction upon the unsuspecting citizens of Lawrence on the morning of August 10, 1863. When the smoke cleared and the sounds of wailing widows subsided, 180 men and boys lay dead in the streets and farm fields around Lawrence.

Almost immediately, Mount Oread was pressed into military service. A battery of cannons were placed atop the high point, and Camp Lookout, also called Fort Ulysses was born. In August 1864, the building of an actual Union fort structure began. The fortified structure was named Fort Ulysses. As the winter of 1864 set in, Fort Ulysses was only partially built and contained a handful of government storehouses. By the end of the Civil War, the fort atop Mount Oread was abandoned. Today, there are no known remnants of the fortifications that called Mount Oread their home.

FORT AUBREY: 1864–1866

Hollywood could not write a life as varied, tormented and intriguing as that of Major Edward W. Wynkoop. Frontiersman, adventurer, sheriff, actor, soldier, Indian agent, humanitarian—Wynkoop performed all these roles, becoming both revered and reviled in the West. Making his way to Kansas Territory in search of wealth and adventure, Wynkoop traversed many Kansas Territorial towns before making his way to Denver. Loyal to the Union and a staunch antislavery advocate, Wynkoop joined the First Colorado Volunteer Regiment. The regiment saw action at the Battle of Glorieta in 1862. After his valiant service, he was promoted to the rank of major and placed in command of Fort Lyon.

Wynkoop and his commander, Colonel John M. Chivington, did not always see eye to eye. A fire-breathing Methodist minister, Chivington held little regard for the Native American population living in Eastern Colorado. Chivington, like many settlers in Colorado, viewed the various Native American tribes, especially the Cheyenne, as nothing more than bloodthirsty savages bent on thievery and destruction of white civilization in the West. Wynkoop initially held similar views. However, after meeting Black Kettle, the famed Cheyenne peace chief, Wynkoop would come to see the Cheyenne peace chiefs and their people as human beings. This view ran counter to those of Chivington.

On September 28, 1864, Cheyenne chiefs Black Kettle and White Antelope met with Colorado governor John Evans and Colonel Chivington. Arranged by Wynkoop, the meeting was intended to ease tensions between the two peoples and birth a constructive dialogue between them to ensure that peace would prevail on the frontier. Sadly for Wynkoop and the Cheyenne, this was not to be.

On November 29, 1864, while the Civil War raged in the east and Confederate general Sterling Price had been defeated in Kansas earlier in the fall, Colonel Chivington and his troopers prepared to mount an attack that would ignite the Plains Indian Wars for decades. The morning was cold. As the Cheyenne slept warm in their lodges along Sand Creek, Chivington recalled, "The morning of the 29th day of November, 1864, finds us before the village of the Indian foe. The first shot is fired by them. The first man who falls is white. No white flag is raised. None of the Indians show signs of peace, but flying to rifle pits already prepared they fight with a desperation unequalled."[64] With over seven hundred cavalrymen as his right hand, Chivington dispensed his own form of plains justice—Black

Kettle and White Antelope and their followers never knew what hit them until it was too late. By the time the day was done, men, women and children lay dead on the cold ground. For Wynkoop, Sand Creek was foul and unjustifiable. Despite his objections to Chivington's methods, he still wore the Union uniform and continued to serve his country.

The relationship between Chivington and Wynkoop had not always been strained. On May 12, 1864, Chivington instructed Wynkoop to "establish a stray pickett down the Arkansas River, say at the southeastern line of Colorado,"[65] to prevent Native American incursion into settled areas and keep an eye out for Confederate activity from the Indian Territory (Oklahoma) that may possibly have spread northward into

Edward W. Wynkoop (left), former Union soldier and Indian agent, was a fixture at many Kansas forts in the nineteenth century. *Courtesy Deb Bisel, author's collection.*

southwestern Kansas and could threaten Colorado. Wynkoop carried out his orders without delay. On May 21, 1864, Wynkoop reported to Chivington from Fort Lyon, "I have just received dispatch from Lieutenant Wilson, Commanding Camp Wynkoop, a picket camp 60 miles east of this post, to the effect that the Cheyenne are about establishing a large camp in this vicinity."[66] Camp Wynkoop had been established just in time, it seemed.

For settlers living in the region, the camp provided a measure of additional security when soldiers occupied the fortification. Robert M. Wright, an early pioneer, moved closer to Camp Wynkoop and built a ranch with a business partner. Wright received various government contracts to haul hay to Fort Lyon and found Camp Wynkoop to be exemplary. From May to September 1864 and August to September 1865, soldiers from various units occupied the outpost. Protecting trade and settlers in the region, the camp became a focal point for travelers and locals alike. The camp was not continually occupied. As the threat of Native American conflict increased, settlers would retreat to the camp's protective walls, regardless of the presence of soldiers. Major General Dodge believed the post to be useful and sent a unit of Galvanized Yankees, the Fifth U.S. Volunteer Infantry, to the camp. On

September 15, 1865, the camp was renamed Fort Aubrey by Special Order No. 20 issued by Major General W.L. Elliott, headquartered at Fort Lyon. No reason was given in the order for the change of name. Regardless of the name, the fort continued to soothe the ragged nerves of those living in the region. Boasting aggregate troop strength of three hundred in October 1865, Fort Aubrey was growing.

Perhaps the best testament to the importance of Fort Aubrey came from those who benefited from its services. While away conducting business between Fort Aubrey and Fort Lyon, Robert Wright did not fear for the safety of his wife and children. He confided to a friend, "My wife and children are there [Fort Aubrey] too, in one of the strongest little forts in the country, six or eight men with them, and plenty of arms and ammunition; all the Indians on the plains could not take them."[67] William N. Byers, traveling east from Denver on a stage, made a stop at Fort Aubrey and felt compelled to write a letter to his newspaper, the *Rocky Mountain News*. On January 7, 1866, he said of Fort Aubrey:

> *We are now forty-five* [i.e., 20] *miles east of the Colorado and Kansas line. This post* [Fort Aubrey] *was established late in the fall, and the work arrested by a heavy snow storm in the first days of December. It is intended to check the Indians in their frequent raids upon the road, and is at present garrisoned by two companies of the 48th Wisconsin— infantry—and one company of the 2nd U.S. Cavalry. The men are quartered in half underground caves, dug and built in the bank of a little spring branch about three hundred yards from the Arkansas River. The only other buildings put up are of adobe or sod covered with earth. Timber is scarce; only a little green cottonwood upon the islands and sand bars in the river. Lumber is brought from Leavenworth after first coming from the pineries of Wisconsin.*[68]

Like many other small outposts on the Kansas plains, Fort Aubrey outlived its usefulness as a protective barrier against Native American depredation with the negotiation of treaties and removal of the tribes to reservation lands. On April 15, 1866, Fort Aubrey was abandoned by the military. The old post continued to serve as a stage stop and eventually was used as a ranch. Homesteaders found the area to be an excellent place to build homes due to the proximity to water. A Hamilton County surveyor noted in 1872 that the homesteaders were fortunate for "the soil is excellent and they think have got the garden spot in Kansas...in another year the railroad will be

This cabin is identified as the Block House from Fort Blair/Baxter, scene of the infamous Baxter Springs Massacre. *Courtesy Michelle M. Martin.*

completed through the valley and settlers will begin to flock in and then will this valley—the Great American Desert of geographies—begin to bloom and to bear fruit."[69] Today the remains of Fort Aubrey are located on private property and are not accessible to the public. Like the homesteaders that came before, the current landowner raises cattle and plants crops at the site of the old fort.

FORT BAXTER: 1862–1866

The memory of combat seldom fades with time. For Dr. W.H. Warner of Girard, Kansas, the sights, smells and sounds of sheer bloody mayhem did not leave his mind until his dying breath. "All was now quiet, like the calm after a furious storm, and we had time to make a list of the causalities [*sic*]. Of the forces at the Springs, eight white soldiers and one colored soldier were killed, and about fifteen were wounded, including one woman, shot through the heel, and a little child shot through the lungs…For an hour or two all was quiet, with the exception of our preparations for another attack, which we momentarily expected. We did not know who our enemy was, nor why he had so suddenly left us; but we fully expected him to return. We afterward learned that the enemy was the notorious Quantrill and his guerrillas, remembered Warner."[70] His firsthand account of the Baxter Springs Massacre paints a chilling portrait of war reaching its apex of cruelty.

Founded in 1862, Fort Baxter (also known as Fort Blair) was ideally situated in the Cherokee Neutral Lands in Southeastern Kansas near the Missouri, Arkansas and Indian Territory borders. This proximity to two Confederate states and unpredictable Indian Territory made Fort Baxter/Blair the ideal location for a military post. The post was also a halfway point between Fort Scott to the north and Fort Gibson in the Indian Territory linked by the Military Road.

The post consisted of several log cabins that were enclosed on four sides by walls roughly four feet in height. Earthen embankments were thrown up against the log walls. Soldiers garrisoned at Fort Baxter enjoyed a life of relative leisure early on during the Civil War. "Here we camp, with nothing to do but eat, drink, swim, sleep and read, the latter only when we are fortunate enough to procure newspapers or books."[71] However, life for soldiers at Fort Baxter/Blair would take a chilling turn for the worse in October 1863.

In October 1863, Lieutenant James Pond was dispatched from Fort Scott to Fort Baxter/Blair to assume command. Roughly 150 men, 75 of whom were African American troopers, defended the post. Upon Pond's arrival, he deemed the post too small to house the troops garrisoned there. Pond ordered the west wall of the fort to be dismantled to extend the boundary of the post to include his tent. He then sent out a foraging party of 60 men, along with all the post's wagons. This left Pond with a paltry force of 100 troopers, including his 75 African Americans.

Driving his men hard, William Clark Quantrill was making his way to Confederate Texas to rest and resupply during the winter months. Capturing a Federal wagon train on the morning of October 5, 1863, Quantrill and his men extracted information from their Federal prisoners and then made their way to Fort Baxter/Blair.

Once the attack upon Fort Baxter/Bair began, chaos ruled. The quick thinking and bravery of Pond in ordering his men to charge through Quantrill's line and enter the fort saved many lives. Upon arriving at the fort, Pond and his force were able to turn a Howitzer on the invaders and

A windswept cemetery west of present-day Baxter Springs, Kansas, is home to a monument commemorating the deaths of Union soldiers killed in the Baxter Springs Massacre in October 1863. In 1871–72, the bodies of the men were disinterred and moved to this cemetery plot and a commemorative marker was erected. *Courtesy Michelle M. Martin.*

fire. With a small force, Pond was able to save the fort and defend his position. Major General James Blunt, however, would not be so lucky.

Saddled with a wagon train and mounted troops, Blunt and his hundred troopers were passing through the area. Just north of the fort, Quantrill's men approached Blunt and his forces. Mistakenly believing them to be an honor guard sent to greet him, Blunt's men paid dearly as the Confederate partisans charged toward them. Organizing a hasty battle line, Blunt's men were scattered. By 5:00 p.m., when it was all over and the smoke and fog of war cleared, eighty-five of Blunt's men were dead. Lieutenant Pond suffered six casualties. The Baxter Springs Massacre would go down in the annals of Kansas military history as a resounding failure. Blunt would be relieved of command after the fiasco, and the post would be abandoned after the war. Today, remnants of the fort are located in the town of Baxter Springs and a modern museum preserves the history of one of the darkest days in Kansas military history. The windswept cemetery west of Baxter Springs, Kansas, is the final resting place for men who perished that awful day defending Fort Baxter.

Fort Brooks: 1864–1865

While the Civil War raged in the East and along the Kansas/Missouri border, the citizens of north-central and northwestern Kansas were vulnerable to attack from various Native American tribes, especially the Cheyenne. The Republican, Solomon and Smoky Hill River valleys were highly prized by settlers looking for fertile farm land and by the indigenous tribes that were used to ranging freely on the lands along the rivers.

In the fall of 1864, the left bank of the Republican River became the site of Fort Brooks, named for George Brooks, a local farmer whose land the post was constructed on. With a log blockhouse, the fort was the local headquarters, charged with defending the area against attack from Native American tribes in the region. Once located near the present-day community of Clyde, Kansas, today nothing of Fort Brooks remains.

Camp Sackett, Kansas Territory, as depicted in *Harper's Weekly*. Camp Sackett was typical of many early Kansas forts and was transitory in nature, with temporary lodging for its inhabitants. *Courtesy Deb Bisel, author's collection.*

FORT BELMONT: 1864–1865

Built to protect the residents of Woodson County from incursions of dreaded border ruffians from Missouri and potential confrontations with Native American tribes, Fort Belmont was more than a military garrison. With several cabins for officers, a redoubt built of earthworks and logs and a parade ground, Fort Belmont would become important to Native American–white relations in Kansas.

Supported by companies C and G of the Kansas Sixteenth Regiment (local militia), the soldiers garrisoned at Fort Belmont actually were local men and many stayed in their homes, near the town of Belmont and the surrounding countryside, more than they stayed at the post. The post saw most of its activity when a Federal agency for the Osage and Creek Tribes was located at the post. This agency distributed supplies to Union-loyal Osage and Creek peoples that were escaping the confusion and tumult in the Indian Territory. Fort Belmont would be one of two forts that housed the followers of the Creek leader Opothleyahola after his fight from the Indian Territory. However, by 1864, the agency was discontinued and Fort Belmont was no longer needed. Standing until 1871, Fort Belmont faded into the pages of Kansas history and never saw prolonged action after 1864.

This collection of seemingly defenseless tents would become Fort Ellsworth/Harker, located in present-day Ellsworth, Kansas. *Courtesy Deb Bisel, author's collection.*

FORT HARKER (FORT ELLSWORTH): 1864–1873

The former blockhouse at Fort Harker. The structure preserves the history of the post and houses an impressive collection of artifacts. *Courtesy Michelle M. Martin.*

As the American Civil War raged in the east, the windswept plains of Kansas were rife with turmoil. To maintain peace on the plains, the United States Army built a series of permanent and temporary fortifications along the Santa Fe and Smoky Hill River Trail networks. Situated on the site of an abandoned ranch, General Samuel R. Curtis sent Company H, Seventh Iowa Cavalry to erect a new military post. Originally called Fort Ellsworth in honor of Colonel Ephraim Elmer Ellsworth, the first Union officer killed in the Civil War, the post was nothing to write home about. Lieutenant Edwin de Courey wrote to his superior officers in December 1865, "The men of the command are suffering very much from Diarrhea and other diseases and as there is no medical officer or medications of any kind at this post, I would respectfully request that a medical officer be ordered here to render medical assistance."[72] As if this weren't bad enough, Private John Morrill remarked in a letter to his wife, "We are now at Fort E as it is termed but you would smile to see the Ft. There is a group of log shanties covered with dirt. Most of the windows are made of boards hung on leather hinges & made to swing open and shut. There is two or three of them which have a half window sash and some of them two or more have glass in them. I suppose the aristocracy reside in them which have glass."[73] Such was the nature of army life in Kansas.

By the spring of 1866, suffering at Fort Ellsworth had reached an all-time high. Fire destroyed the post's hay supply. Post commander Captain Kilburn Knox declared the buildings of the post wretched, deplorable and uninhabitable. By the spring of 1866, construction had begun on new post buildings that would be located a mile northeast of the original site. When Alice Blackwood Baldwin arrived in 1867, the new Fort Ellsworth was still under construction. "When I entered my new abode I gazed with disgusted disappointment around the bare squalid room. Its conveniences were limited to one camp chair, two empty candle boxes and a huge box stove…My house contained two rooms—the aforesaid kitchen and "drawing room," one end partitioned off by a grey army blanket. Behind this barrier was the sole bedroom accommodations of the dug out."[71] For a woman, this kind of existence was nearly unbearable. In addition to substandard housing, the inhabitants of the post were dealt a bitter blow by Mother Nature. The winter of 1867 went down in the record books as exceptionally bitter. Predators like wolves and coyotes were on the move in search of food due to the extreme cold. For Alice Baldwin, the sounds of wolves howling and clawing at doors in search of scraps of food must have been terrifying.

Constructed by soldiers and civilians, the new post must have seemed like a posh hotel suite to those that endured the original quarters. With a hospital, officer's quarters, storehouses, stables, corrals and an icehouse, the post was complete. The newly reconstructed post was rechristened Fort Harker in honor of Brigadier General Charles Garrison Harker, who was killed in action at the battle of Kenesaw Mountain, Georgia, during the Civil War. While stationed briefly at Fort Harker, Captain Albert Barnitz wrote to his wife, Jennie, that he was "much pleased with Fort Harker, in spite of all the disagreeable surroundings. The horses are in good stables… the officer's quarters are progressing finely and they will be handsome indeed—even more pleasant and cozy than those at Fort Riley—and they are beautifully situated."[75] Once complete, the post boasted several structures situated around a main parade ground. With adequate housing and support structures, the post could now turn to the business of protecting the trails.

From 1866 to 1867 Fort Harker busied itself with completing the construction of the post and assisting in the delicate work of maintaining Native American–white relations in the West. With Indian agent Edward W. Wynkoop at Fort Harker, the post was a hub of activity. The year 1867 would prove to be a watershed year in Fort Harker's term of service on the plains. When the Seventh Cavalry moved its headquarters to Fort Harker in 1867, activity on post increased. Troopers, officers, officers' wives, families and visiting troops made Fort Harker buzz with life. General Winfield Scott Hancock used Fort Harker as a resupply station for this winter campaign against the hostiles in the region. With conflict between U.S. troops and Native Americans on the rise, soldiers from Fort Harker provided escort service and special guard duties over 190 times as they accompanied supply trains along the trails.[76] Fort Harker would train and muster troops into service in 1867 and protect settlers and wagon trains from increased hostility on the part of many tribes, including the Cheyenne. The Thirty-eighth Infantry, composed of African American troops, would call Fort Harker its headquarters from 1867 to 1869 and performed the majority of the trail escort missions from the post.

As the years passed, Fort Harker was beset with hardship. A cholera epidemic in the summer of 1867 wreaked havoc on the post and surrounding countryside. Nearly one hundred soldiers and civilians died at Fort Harker alone from the disease, with the living often times too ill to bury the dead. By 1870, Fort Harker, along with many of its counterparts in Kansas, was deemed obsolete. After serving as a bulwark against Native American attack, a supply post and operating base for army operations, Fort Harker was shuttered for good on April 2, 1872. Located in Kannopolis, Kansas, the remains of the post—a junior officer's quarters, officer's quarters and blockhouse—are museum buildings run by the Ellsworth County Historical Society. Again, visitors can walk in the shadows of such historical figures as Custer, Sheridan, Hancock, Wynkoop and Elliot.

FORT LINCOLN: 1861–1864

With Kansas's entry into the Union as a free state on January 29, 1861, James Henry Lane would become one of Kansas's first U.S. senators. This position afforded Lane the access to the political power he craved. By the summer of 1861, Lane had been commissioned a general in the United States Army and was operating in Kansas once again. Charged with raising regiments of African American and Native American troopers for the Union cause in the West, Lane was an active participant in Kansas military affairs.

In the summer of 1861, Lane established a fort on the north side of the Osage River in Bourbon County west of the present-day community of Fulton. Lane named the post Fort Lincoln in honor of the president and erected a blockhouse surrounded by a stockade. Fort Lincoln served as a part of the border defenses built to protect Kansas from invading Confederate forces from Missouri to the east. The fort also became a prisoner of war camp. Troops would be garrisoned there under Lane's command until 1864, when the buildings were moved to Fort Scott. Today, a one-room schoolhouse that was built near the fort stands on the campus of Fort Scott Community College, a reminder of the perilous days of war.

FORT LOOKOUT: 1861–1869

With no stockade, stables or officers' quarters, many Kansas "forts" established to police the various trails and military roads were simple structures. Providing crude shelter and minimal defense, many military posts in western Kansas were transient in nature and not deemed worthy of finery found at Forts Leavenworth, Riley or Larned. Established in 1861 to guard the transport of goods, men and materiel along the Fort Riley to Fort Kearney military road, Fort Lookout (also called Lookout Station) was perched atop a high bluff overlooking the Republican River in Republic County just two miles from the Nebraska border. Two stories in height, something unique for satellite posts, Fort Lookout was built on a forty-five-degree angle and boasted eight sides. This provided the men posted at the fort a well-rounded view of the countryside. Built with sturdy, thick logs, the two-story post also featured several notches cut out of the second story logs that served as defensive gun ports in the event of attack.

As the war drew to a close, the threat of Native American attack subsided and Fort Lookout was no longer needed to guard the military

This photograph from 1909 shows the battered remains of Fort Lookout. *Courtesy Deb Bisel, author's collection.*

road. Abandoned by the army by 1868, the old post would be utilized by local militia units from 1868 to 1870 for protection during various Native American raids and uprisings. On May 26, 1868, a hunting party consisting of six men and two wagons hauling buffalo meat began their return trip to Waterville, Kansas. Heavily loaded with meat and hides, the group moved slowly. Seeing Native Americans atop a ridgeline, the group made a breakneck dash for Fort Lookout. Just as settlers in the area had come to rely on seeking refuge behind its thick walls, travelers still depended on the abandoned fort. Running low on ammunition, the party reached the fort and believed themselves safe. By morning light, they believed their pursuers had left during the night, but as they left the protective walls of the fort, still low on ammunition, the men faced an awful fate. Attack came from all sides quickly. Of the entire party, only one survived. Today the graves of six men lost and Fort Lookout have all vanished from the Kansas landscape.

FORT ROW: 1861–1865

Constructed after Confederate forces burned Humboldt, the Union post Fort Row was charged with protecting the citizens of Coyville and the surrounding countryside near present-day Fredonia. Named for Captain John R. Row, the fort was built in the fall of 1861 "at a point about three miles south of the town, on land now owned by John Shaffer. This was named Fort Row, in honor of the captain of the force. It consisted of three block houses, 16x24 feet, made of heavy logs, and enclosed with pickets six feet high. An embankment was thrown up on all sides, and the company went into winter quarters."[77] With approximately eighty men, Fort Row was charged with protecting settlers, assisting other Kansas militia units and maintaining the peace during the Civil War. By the spring of 1862, the majority of the men serving at Fort Row enlisted in the Ninth Kansas Volunteer Cavalry, and the fort was all but abandoned. Before the fort fell into disuse, it became the scene of a humanitarian crisis the scale of which no one living in the area could comprehend.

Hungry. Cold. Exhausted. Terrified. Wintry January winds blew as the refugees trickled in ten to twenty at a time until nearly ten thousand men, women and children had made the dangerous trek from the Indian Territory to the small military post. With hardly enough supplies for the men at the post, Fort Row was not prepared to feed and clothe the followers of the

This unidentified American Indian Union soldier was a member of an Indian Home Guard unit mustered into service in Fort Scott, Kansas. After fleeing from Indian Territory to Fort Row and Fort Belmont, many Creek men joined Indian Home Guard Units and were pressed into service in the homeland they had fled. Unknown Native Soldier 30114 in the collection of Wilson's Creek National Battlefield. *Image courtesy of the National Park Service.*

Creek leader Opothleyahola after their flight from Indian Territory. Dissent amongst the various tribes living in Indian Territory, a lack of early Union support during the war and the Union's abandonment of Forts Arbuckle, Washita and Cobb in 1861 created a power vacuum. For the pro-Confederate tribal factions, this was a prime opportunity to seize control and prove their loyalty to Richmond. Not all tribal leaders believed Confederate promises, and old animosities flared among tribal factions in the Creek nation.

Turning against those Creeks who supported the Confederacy, Opothleyahola was branded a renegade and enemy by his own people. Creeks who wished to remain neutral followed his leadership. Doggedly pursued by Confederate troops through Indian Territory, Opothleyahola and his followers fought a series of running battles as they headed north to the safety of Union-controlled Kansas. In November 1861 at Round Mountain and in December at Chusto Talasah and Chustenalah, Opothleyahola's troops fought against Confederate Creek and Cherokee warriors. By December 1861, what was left of Opothleyahola's followers walked to Fort Row. When they arrived, they discovered little food and water and few blankets or supplies. Within two months of their arrival, over 250 Creeks perished; hundreds of frostbitten, frozen limbs were amputated in order to save their owners. Many of the beleaguered continued to walk until reaching Fort Belmont in search of sustenance. Opothleyahola would not live to see the war's end, dying in Kansas in 1863. His trek for freedom is commemorated along Kansas highways that traverse his route. Nothing remains of Fort Row.

FORT SIMPLE: 1864

Confederate major general Sterling Price was a big man. More than six feet tall and more than three hundred pounds, he was also handsome with a mane of wavy, white hair. He was a combat veteran, having seen service in the United States Army during the Mexican War. When news reached Topeka that he was marching toward the Kansas border, concern—if not outright fear—gripped the town. It was little more than a year since Confederate guerrilla leader William Clarke Quantrill had destroyed the town of Lawrence and killed 150 men and boys in the process. Prior to that, most people thought they was too far inside the state line to be the target of a raid. Now, everything was uncertain. Kansans in the 1860s realized nothing was off the table.

The news that Price was steadily moving forward prompted Governor Thomas Carney to call out the state militia, declaring, "The State is in peril!...Men of Kansas, rally!...To arms, then, To arms and the tented field, until the rebel foe shall be baffled and beaten back."[78]

According to historian Fry Giles, the Second Regiment of the Kansas militia, mostly Shawnee County residents, immediately mustered in the capital city and was sent to the Missouri border. Commanded by Colonel George W. Veale, there were 561 men and two cannon. Those locals who did not immediately arrive in Topeka were organized under Major Andrew Stark and formed a home battalion. The battalion included a company of militia commanded by Giles himself and a captain, as well as two companies of cavalry and a company of "colored recruits, of sixty-five men, under Capt. Thos. Archer." The battery section had only one gun, and there was a company of infantry and one of "exempts."[79]

To protect the capitol, a stockade was built at the intersection of Kansas Avenue and Sixth Street, on the east side of the city. Trenches were dug a few blocks east of that location at Sixth and Jefferson and Eighth and Madison.

"For the period of two weeks, during which these men were under arms, they did the best they could to make of themselves soldiers; and to realize their anxiety in this regard," recalled Giles, "it should be remembered that it was the popular impression that if the rebel army succeeded against the force that had gone to the border to oppose it, that army would be at once broken into detachments, for raiding upon the towns of Kansas, as the most effectual method to punish that people who, in the estimation, of the rebels, were beyond all others to be blamed for the war."[80]

When Price was repelled, Topeka breathed a sigh of relief, but it would be five or six years before Fort Simple, built to defend the capital of Kansas, was dismantled.

FORT SOLOMON: 1864–1865

Originally known as the Nepaholla, the Solomon River is one of the major waterways in northern Kansas.[81] The Solomon River attracted settlers as early as 1854 with its fertile lands. Life was harsh for those first pioneers that settled the lands along the Solomon River. Fighting grasshoppers, drought and intermittent attacks from Native American tribes in the region, settlers banded together for protection. With the onset of the Civil War, settlers endured not only the loss and privation that war wrought upon them but also the heightened danger of attack from the Cheyenne. By 1864, they had begun construction on a small outpost that was home to several log structures surrounded by a protective palisade. To assist the settlers in their defense, Major General Samuel R. Curtis left the defenders of Fort Solomon "Jim Lane's Pocket Piece," a small artillery piece.[82] As with many of these hastily constructed frontier outposts, Fort Solomon was abandoned with the close of the war, and the Ottawa County countryside was once again peaceful.

FORT ZARAH: 1864–1869

The attack was unexpected and the troopers unprepared. Lulled into a false sense of security by the sight of Union uniforms, General James Blunt and his men never imagined the men that galloped toward them would be their undoing. When the shots rang out, the surprised Union troops scrambled to return fire. With the fog of war settling around them, Major H. Zarah Curtis fought until unhorsed. Taken prisoner, the son of General Samuel

R. Curtis died at the hands of William Clark Quantrill's men at the Baxter Springs Massacre in October 1863. By all newspaper accounts, the funeral procession of Major Curtis was the largest ever seen in Keokuk, Iowa.

By 1864, General Samuel R. Curtis was in command of Union forces in Kansas. Charged with protecting Kansas from Confederate invasion, Curtis also had to face the threat of Native American attacks in the western regions of the state. To aid in the protection of Western Kansas, a camp was established thirty miles east of Fort Larned along Walnut Creek. "The site of the fort is well known to all at the southeast of the Walnut Creek. It stood on a gentle eminence, nearly surrounded by a shallow creek which might easily be made a means of defense by filling it with water."[83] Constructed of earthen dugouts and tents on a high creek bank, the post received its orders from Fort Larned's commanding officer initially. Called Camp Dunlap originally, Curtis renamed the outpost Fort Zarah in honor of his slain son.

Major H. Zarah Curtis, the namesake of Fort Zarah, lost his life in the Baxter Springs Massacre in 1863. His father, General Samuel R. Curtis, named the post in his son's honor. Major Henry Zarah Curtis 30203 in the collection of Wilson's Creek National Battlefield. *Image courtesy of the National Park Service.*

With the Civil War drawing to a conclusion in the East, the Plains Indian Wars in the West reawakened. To reinforce the troops there, the Second Colorado Cavalry was stationed at Fort Zarah and commenced construction on a limestone octagonal blockhouse. With five square miles of land placed under its jurisdiction, Fort Zarah was declared an independent post and administered its own affairs. Assisting in the defense of settlements against Indian attack after the Civil War, Fort Zarah was reinforced with defensive buildings made of locally quarried sandstone. By 1869, however, the post was declared obsolete. In his final report, post commander Captain Nicholas Nolan noted that "the roof is of tin and very valuable and the flooring has been all placed within two years, and are almost new, and both can be used with the windows again, without any expense except transportation and the

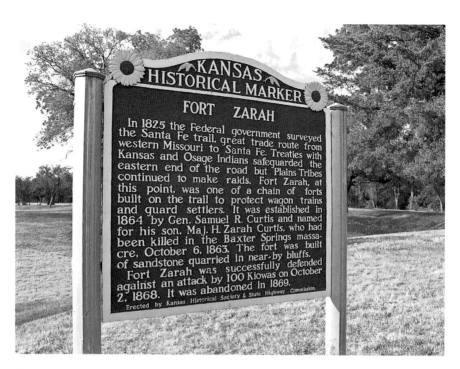

This historical marker denotes where a portion of Fort Zarah was located three miles east of present-day Great Bend, Kansas. *Courtesy Michelle M. Martin.*

necessary mechanical aid which I can supply from my troops; and in my estimation if the Post be left in status quo it will become a rendezvous for horse thieves and prairie robbers."[84] Today, a roadside park three miles east of Garden City, Kansas preserves the history of Fort Zarah and commemorates its role in defending the Kansas frontier.

Chapter 3

THE SOLDIER STATE: THE INDIAN WARS AND WESTWARD EXPANSION

1865–1899

The houses here are more picturesque than <u>elegant</u>. I wish you could see our mud house. It looks gloomy now but when our boxes come I will fix it nicely. When our new quarters are done next year, we will live somewhat differently. Now we are living somewhat in the manner of Prairie dogs. The men's quarters are holes burrowed in the bank of the river.
—*letter from Mrs. Isadore Douglas (wife of Major Henry Douglas, commander of Fort Dodge) to her mother, February 3, 1867*

As the American Civil War came to a dramatic conclusion with the assassination of President Abraham Lincoln, the Untied States Army began the process of Reconstruction in the former Confederate States of America. Taking advantage of the Homestead Act of 1862 and the promise of prosperity and a fresh start, thousands of settlers streamed westward. So many former Civil War veterans settled in Kansas that it was given the moniker "the soldier state." Seeking peace, many of these settlers would once again find war, this time with an enemy that did not fight like regular army. For the Native American nations in the plains, settlers meant encroachment onto lands they had called home since time immemorial. Conflict between settlers and Native American tribes created the need for military forts and the protection soldiers could provide. Kansas was home to numerous posts that played an integral role in the military actions during the Plains Indian Wars. At these posts served some of the most storied names in American military history: Custer, Cody, Hickok, Sheridan, Forsyth, Wynkoop and

Wild Bill Hickok and Buffalo Bill Cody captured the American imagination during their lifetimes. Hollywood embraced the deeds and legends of the two men on celluloid. In the Paramount Pictures 1936 film *The Plainsman*, Buffalo Bill was portrayed by James Ellison, and Gary Cooper played Hickok. *Courtesy Deb Bisel, author's collection.*

Harper's Weekly filled its pages with words and images of the West for an eastern audience that clamored for heroic stories. Here, coyotes threaten a tent at night. *Courtesy Deb Bisel, author's collection.*

Hancock. Kansas became a proving ground for some of America's most famous Indian fighters. Native American leaders would also gain notoriety. Roman Nose, Tall Bull, Dull Knife, Bear Bull, Satanta and Black Kettle all traversed the wide-open Kansas landscapes. Kansas once again was the epicenter of conflict and change.

CAMP BEECHER (CAMP BUTTERFIELD, CAMP DAVIDSON): 1868–1869

Life on the Kansas plains in the nineteenth century was difficult at best. With the close of the American Civil War, settlers came streaming into Kansas. The lure of wide-open spaces and land for agriculture appealed to those that had lost much during the war. In 1862, President Abraham Lincoln signed into law the Homestead Act. This one piece of land-use policy did more to change the nature of the central plains than any other legislation before or after. Settlers now had a legal vehicle by which they could settle tracts of land in an affordable manner. As the settlement floodgates opened, a tidal wave of change and conflict swept across the Kansas prairie.

Kansas governor Samuel J. Crawford appealed to General Phillip H. Sheridan, of the Department of the Missouri at Fort Leavenworth, to establish a permanent post in south central Kansas for the protection of settlers and trade in the region. Given the proximity of Butler and Segdwick Counties to Indian Territory to the south, the request made perfect sense to Crawford. On April 14, 1868, Sheridan replied, stating, "I am in receipt of your letter in reference to the establishment of a military post at the mouth of the Little Arkansas, to protect the settlers in the county of Segdwick." He further stated that he intended to "send a small military force there and the place will be occupied by at least one company by the 1st May." First called Camp Butterfield then renamed Camp Davidson, the post was established on May 11, 1868, at the junction of the Big and Little Arkansas Rivers in present-day Wichita. On October 19, 1868, the camp was renamed in honor of Lieutenant Frederick H. Beecher, the much-lauded martyr of the Battle of Beecher's Island.

Camp Beecher's founding was fortuitous, for on May 17, 1868, newspapers in the region ran headlines of sheer bloody murder. The *Kansas Daily Correspondent* of May 27, 1868, heralded "Two Men Killed by Osage Indians On Big Walnut, Butler County, The Bodies Frightfully Mutilated and the

Tribe Surrenders Two of the Guilty Parties for Trial."[85] Nervous settlers in the region clamored for protection. As with many news articles of the time period, the more salacious and gruesome the detail the better for a reading public to devour. In describing the killings of Sam Dunn and James Anderson, settlers living on the government strip, the correspondents spared no expense: "Their heads were cut off and scalped, that of the former being left several rods from the body. The fingers were also cut off from one of the bodies and taken away. After the massacre was completed, the party let down the fence to an eighty-acre field nearby, and drove off two mules; they also chased a horse to within a hundred and fifty yards of the house." The arrival of the troops was opportune to quell the rising tide of fear among settlers.

Short on supplies from the Department of the Missouri, the men first sent to Camp Beecher relied on elbow grease and spades to dig their quarters out of the earth. Located near the Little Arkansas River, the men could fish and pick wild sand plums and elderberries. The number of troops stationed at Camp Beecher fluctuated. By 1868, the adjutant general's report indicated that two companies of the Seventh U.S. Cavalry and one company of the Fifth U.S. Infantry were stationed at the little post. With a chaplain, an adjutant and one regimental quartermaster, the post hummed along efficiently.[86] Captain Robert M. West proved to be Camp Beecher's most well-trained and venerated commander. A veteran of the Civil War, West participated in the Peninsular Campaign and commanded Fort Magruder. Honorably mustered out of service in 1865 with the rank of brevet brigadier

Major General Frederick Funston from Iola, Kansas, was the son of a Civil War veteran, the father of a World War II veteran and a Medal of Honor recipient. When he died suddenly in Texas in 1917, he lay in state at the Alamo (the first person so honored) before his remains were sent to San Francisco, where he lay in state in the rotunda of the city hall. He was buried in full dress uniform at the Presidio. *Courtesy Kansas Museum of the National Guard.*

general, West would serve as a captain of the famed Seventh U.S. Cavalry in the West. Three other men—Captain Samuel L. Barr, First Lieutenant George McDermott and Captain Owen Hale—all made the tiny outpost their home for a brief period of time.

In June 1869, troops of the Seventh U.S. Cavalry stationed at Camp Beecher left the post for the final time. As reported by General Schofield, commanding Department of the Missouri, the "Seventh cavalry, now at Camp Beecher, will at once move northward towards the big bend of Smoky Hill, scouring the country between the Arkansas and Smoky Hill. If no Indians are discovered they will go to Fort Harker."[87] The Seventh never returned to Camp Beecher. Today, the modern city of Wichita, Kansas, envelops the location of old Camp Beecher.

CAMP CRAWFORD: 1868

As hostilities with Native Americans increased following the Civil War, the Seventh U.S. Cavalry and the Eighteenth Kansas Cavalry found their numbers insufficient to handle the conflicts. General William T. Sherman authorized the formation of another regiment of Kansas Cavalry and General Phillip Sheridan wrote to Governor Sam Crawford requesting those men in October 1868. The fiery governor did not hesitate to comply. He had traveled to the western part of Kansas to discern for himself the real situation. He found "men, women and children were murdered indiscriminately…the country laid in ashes and the soil drenched in blood."[88]

Men immediately answered the call and began arriving in Topeka, where a camp was established on the Kansas River on the edge of town to muster them into service. The men named the camp in honor of the governor, who shortly resigned the governorship to be its colonel. The swelling number of soldiers soon had an impact on the capital city. Alfred L. Runyon was one of the first to enlist in Manhattan and wrote from the camp in November 1868:

> *The Manhattan boys arrived in Topeka about noon, safe and sound, on Monday last. At Wamego we were joined by another party of recruits bound for the same destination.*
>
> *We marched into Camp Crawford, and pitched our tents. The wind blew very hard, causing a great dust, which did not increase the good humor of the men.*

On Monday evening a lot of horses stampeded from the corral, situated about half a mile from camp, and made a terrible clattering going over the bridge. They were recovered the next morning, except about eighty. The same evening there was a row in a house of ill fame in the city, during which one of the soldiers belonging to camp was badly wounded. On Tuesday evening a man named Williams, of Company "D" was shot in the side by an accidental discharge of a musket, in the hands of one of the guards. Luckily the shot glanced and inflicted only a slight flesh-wound. Gen. Sheridan was in town Tuesday morning, but left in the afternoon.[89]

The Nineteenth was a part of the winter campaign that included the attack on the Washita in the Indian Territory. While the regiment did not arrive in time to take part in that engagement, Colonel Crawford was with George Custer when he returned to the frozen battlefield days later to find the bodies of Joel Elliott and his men, left behind by Custer's command.

CAMP KIRWIN: 1865–1865

Some boys want to be doctors or lawyers when they grow into manhood, but not John S. Kirwin. Born in New Hampshire to Irish immigrant parents, young Kirwin read the novel *Charles O'Malley, or The Irish Dragoon* at a fairly young age. "This fixed my ambition to become a soldier, when I was old enough to become one."[90] In 1858, Kirwin, age eighteen, ran away from home to escape a life of farming and traded the plow for a rifle. After enlisting in the regular army in Boston, he was sent to Carlisle Barracks, where he attended the school of instruction for cavalry. His dream was coming to fruition. Before he would be tested in battle in the American Civil War with the Twelfth Regiment Tennessee Volunteer Cavalry, Kirwin headed west in 1859 to his new duty station—Fort Riley, Kansas Territory.

Serving in the Fourth Cavalry Company K, Kirwin recalled many of the individuals that he served alongside, including J.E.B. Stuart and Major James Longstreet. Itching for action, Kirwin's wish was granted in the fall of 1859. The summer of 1859 saw increased traffic along the Santa Fe and Smoky Hill Trails with the discovery of gold at Pike's Peak in the western fringe of Kansas Territory (later Colorado). Leaving Fort Riley in the summer of 1859, Kirwin and the men of the Fourth Cavalry provided protection for travelers and settlers in the Arkansas River valley from attack from the Kiowa, Comanche,

Cheyenne, Arapaho and Apache. After seeing little action, they began their return trip to Fort Riley in September. Kirwin would later recall, "We reached the Little Arkansas River on the evening of September 30th. About 2:00 AM the 1st Sergeant laid his hand on my shoulder and whispered in my ear 'Get up quick and make no noise, the Indians have broken loose and killed Peacock and burned his ranch'; this ranch was thirty miles back on the road we just traveled."[91] Returning to the ranch, Kirwin and his comrades found the home ablaze and the unfortunate owner indeed scalped. This would not be the first time Kirwin would experience death.

From 1861 to 1865, Kirwin led the Twelfth Regiment Tennessee Volunteer Cavalry as its colonel. Little did he know, in May 1865 his unit would be ordered to report to St. Louis and then to Kansas to assist in putting down Native American insurrection on the plains. Kirwin, having already served in Kansas, would be a tremendous asset. On November 1, 1865, Major General Grenville M. Dodge, in correspondence with his superiors in the Department of the Missouri in St. Louis, proposed a military strategy against the tribes in Kansas. Dodge stated, "In forming my plans for the campaign, my understanding was that the hostile Indians were to be punished at all hazards, and this I intend to do, knowing if I was allowed to press the campaign according to my plans that before another spring a satisfactory and durable peace could be obtained."[92] Dodge put Kirwin in command of a column of 350 troopers that traveled up the Republican and Smoky Hill Forks of the Kansas River. Their mission: keep the country between the Platte and Arkansas free from Native Americans. In order to accomplish their mission, the men would need a forward base of operations.

Kirwin would have his men erect a stockade on the north fork of the Solomon River that they christened Camp Kirwin. A summer transient camp, it served its purpose of providing protection and shelter for Kirwin and his men. Short-lived and temporary, the camp was of no historical importance as the summer of 1865 was peaceful in the region. The camp, once abandoned, became a protective shelter for travelers and eventual settlers that came in the early 1870s to settle the area. In honor of the man who built the camp, the town of Kirwin was founded across the river from the site of the old campsite. The campsite would have faded into the Kansas countryside and become the stuff of local legend were it not for the building of a dam in the area in 1946. As the Kirwin Dam was constructed, remnants of a small military outpost were discovered and archaeologists from the Smithsonian Institution confirmed the discovery. Today, nothing remains of Camp Kirwin save its name and the memories of John S. Kirwin.

CAMP LEEDY: 1898

The nineteenth century was one of upheaval for one of America's closest neighbors, Cuba. Revolution and insurrection followed one another as the islanders pressed for freedom from Spain. Many Americans supported the country's efforts to be free of an oppressive government halfway around the world. An insurrection began in 1895, and Spain seemed helpless to quell the problem. In 1898, the United States would be forced into the conflict when a U.S. battleship was blown up in Havana harbor. There were 266 marines killed, and "Remember the *Maine*" became a rallying cry as Americans enlisted to avenge the deaths of these fallen servicemen. In April, President McKinley called for 125,000 volunteers, and Congress declared a state of war had existed since April 21. Kansas met the news, and the call, enthusiastically. At the soldiers' home in Leavenworth, there was a parade of 2,000 soldiers in the rain singing and cheering. Spanish flags were burned. Throughout the state, there were salutes and demonstrations in support of the action.

The first of three infantry regiments raised in Kansas was the Twentieth (since nineteen regiments had been raised during the Civil War). Governor Leedy issued the call on April 26, and the Twentieth was mustered in Topeka from May 9 to 13 at the hastily constructed Camp Leedy. The camp was named for the governor and was just blocks south of the capitol on the fairgrounds. The Twentieth was commanded by Colonel Frederick Funston and was soon ordered to San Francisco. The men who came to Topeka to enlist wore expendable clothing, assuming they would be issued uniforms quickly. Not only did they not get uniforms, but they were also living in tents in chilly, muddy conditions. They were stationed in San Francisco without much more and became something of a curiosity in the city.

"[The] public learned of all the queer points in which the Jayhawkers excelled," wrote a historian. "The Kansas camp came to be visited by the idle and the curious as if it were a menagerie of unique specimens. The Kansas boys were quick to 'catch on,' and the crowd looking for strange sights never failed to find a plenty of them. One Kansan attracted great attention by the curious manner in which he ate broth with his fingers."[93]

For men who had grown up far from the ocean, San Francisco's geography offered endless entertainment, and their reactions provided amusement for the Californians

"Many of these men from the prairie never saw a respectable mountain until they crossed the Rockies," said the *San Francisco Chronicle*, "and were never in sight or smell of tide water before. The belated Kansans slept at Sixteenth street station Friday night. When they went to sleep the waters of the bay were lapping the rocks of the embankment. When they awoke the tide was out and there was a wide expanse of mud. The commanding officer called to a sentry, 'Hello! Where's all that water that was out there last night?' 'Darned if I know,' responded the equally puzzled sentry."[94]

It was the consensus, however, that these men appeared willing to fight, and their reputation was proven in combat when they were sent to the Phillipines.

The Twenty-first and Twenty-second were mustered almost simultaneously, also at Camp Leedy. The regiments never left stateside and regretted they did not get the opportunity to fight. The Twenty-third infantry was raised in July and was composed primarily of "colored men." The Twenty-third was sent to Santiago, Cuba, where it remained until February 1899.[95]

The Twenty-third Regiment was something of an anomaly. Commanded by a black officer, Lieutenant Colonel James Beck, they existed only because they had pressed Governor Leedy to allow them to form a regiment. There had been black soldiers since the Civil War, but they were in segregated units. The governor saw the political potential in agreeing to the request. He was running for reelection. The reaction of the black community was positive, though not without some pointed criticism of their own situation in America. "While [black] newspaper editors proclaimed their patriotism, and were in favor of the liberation of black Cubans from the Spaniards," wrote one historian, "there was regret that more was not being done to protect blacks at home. 'The Negroes of this country, while they are ready to fight the battles of Uncle Sam, are nevertheless, seriously under the impression that there are Spaniards nearer home than Spain or Cuba.'"[96]

Camp Leedy existed only long enough for these units to be raised and sent elsewhere. Most of the buildings used for the camp were previous structures and some remain scattered throughout downtown Topeka.

CARLYLE, CASTLE ROCK CREEK AND CHALK BLUFFS STAGE STATIONS: 1860s–1870s

Jutting skyward toward the blue Kansas heavens, craggy, gnarled limestone formations from the Cretaceous period have attracted settlers for generations. Their prehistoric appearance harkens to a time when Kansas was part of

Breathtaking and grand Castle Rock was the location of one of many stage stops that also housed troops along the Butterfield Overland Dispatch. *Courtesy Michelle M. Martin.*

a vast inland sea. As the waters receded, they left behind Niobrara and Dakota limestone and chalk formations that could be seen for miles on the flat treeless Kansas prairie. Today, tourists willing to drive off the beaten path can marvel at nature's handiwork in the Smoky Hill Valley just south of Interstate 70 in western Kansas.

In the nineteenth century, the rock formations were natural places for travelers to stop and gather as they provided shelter and a visual point of reference on the otherwise monotonous Kansas landscape. For enterprising businessman David A. Butterfield, the Smoky Hill Trail and its natural landmarks would mean big business. Forming the Butterfield Overland Dispatch, Butterfield hired a surveying party that included Lieutenant Julian R. Fitch to survey the best route to transport passengers and goods from Kansas to points farther west. In his reports, Fitch noted the landscape and its rugged beauty: "Nine and one-fourth miles west (from Downer Station) we crossed Rock Castle Creek. Camped two days to rest. The scenery here is really grand. One mile south is a lofty calcareous limestone bluff having the appearance of an old English Castle with pillars and avenues traversing it in every direction. We named it Castle Rock."[97] This would be the first of many descriptive references to Castle Rock, a towering limestone formation that attracted wagons trains, settlers, soldiers and Native Americans to its foundations.

Along the trail that Butterfield had surveyed, various station stops were built and manned to tend to the stagecoaches and wagons that would haul passengers and freight for the Butterfield Overland Dispatch. Each station stop had a particular function based on its size and location. Home stations provided food and respite for travelers and those conducting freight along the trail. Feeding and watering both humans and work animals was the main function of a home stop. As traffic on the trail increased, home stops became natural places to garrison troops that were given the task of protecting commerce along the trail. Monument Station, Fossil Creek Station, Chalk Bluffs and Castle Rock all housed troops or local militia units formed for the protection of settlers and commerce in the region.

From 1865 to 1870, the Smoky Hill Trail brimmed with commercial activity. This drew the ire of many Native American tribes that resented the incursion of white settlers and all that followed them into lands that the tribes believed were their domains. The Cheyenne, in particular, wreaked havoc on trail commerce and homesteaders in this region. One only needs

This image from *Harper's Weekly* depicting Native American Ghost Dancers aided in creating a sense of panic and fear on the plains amongst nervous white settlers and army soldiers. *Courtesy Deb Bisel, author's collection.*

to read the nineteenth-century newspaper headlines that heralded each new Native American depredation to understand the tense nature of life in the region. As the days of the trail drew to a close with the completion of more rail lines and the negotiation of treaties with Native Americans that pushed them out of the region, the need for station stops ceased. Many of the station stops became private property, homes, stores and businesses for intrepid homesteaders who wished to carve out a piece of the American dream from the Kansas landscape.

CIMARRON REDOUBT (DEEP HOLE REDOUBT): 1870–1873

The Battle of Poltava. The battle of Bunker Hill. The Charge of the Light Brigade. The Battle of Vicksburg. What links these seemingly unrelated events in military history? They all involved the use of redoubts. Meaning a place of retreat, a redoubt is a hastily constructed defensive position to protect soldiers outside of a main fortification. Somewhere in the lines of communication in western Kansas during the Plains Indian Wars, someone forgot to tell the men who built Cimarron Redoubt (also known as Deep Hole Redoubt) that being near a fort was a good idea.

Southwestern Kansas feared one force more than the harshness of nature, disease or death—the Comanche. From 1867 to 1875 portions of Southwest Kansas were embroiled in conflict with the Comanche. To protect trade along the Fort Dodge to Fort Supply Trail, soldiers from the United States Army built two redoubts on the north and south banks of the Cimarron River. Built by soldiers commanded by Captain John Page, the Cimmaron redoubt was square in shape, measuring sixty feet on each side.

Life in a deep hole in the ground in southwestern Kansas was hardly the posting an army wife desired in the nineteenth century. With the comforts of Forts Leavenworth, Riley, Hays, Harker and Larned, traveling to Cimarron Redoubt was hardly inspiring. Frances Marie Antoinette Mack Roe, the wife of Lieutenant Fayette Washington Roe, wrote in 1873, "The redoubt is made of gunny sacks filled with sand, and it is built on the principle of a permanent fortification in miniature, with bastions, flanks, curtains and ditch, and has two pieces of artillery. The parapet is about ten feet high, upon the top of with a sentry walks all the time. This is technically correct for Faye has just explained it all to me, so I could tell you about our castle on the plains."[98] What Cimarron Redoubt lacked in support from Fort Dodge or Fort Supply it obviously more than made up for in fortification structure and firepower. Roe described her living conditions in her letters to friends and family. "We have two rooms for our own use, and these are partitioned off with vertical logs in one corner of the fortification, and our only roof is of canvass."[99]

Along with its companion located to the north along the headwaters of Bear Creek, Cimarron Redoubt provided protection and respite for mail coaches and military supply and mail wagon trains traveling from Fort Dodge to Fort Supply in the Indian Territory. By 1873, as hostilities with the Comanche waned, the need for the redoubts decreased. In 1876, the former military fortification was used by the settlers of Clark County as a post office and general store. By 1887, however, the redoubt was abandoned and nature reclaimed it for herself. Located entirely on private property south of present-day Ashland, Kansas, the Cimarron Redoubt was placed on the National Register of Historic Places in 1978.

DOWNER STATION (FORT DOWNER): 1863–1868

Gold and all that glittered attracted prospectors, settlers, gamblers, hustlers, cons and whores to move to the western extremes of Kansas Territory (present-day Colorado) in the late 1850s and early 1860s. With Pike's Peak setting off a new epidemic of gold fever, prospective fortune hunters outfitted themselves in Missouri and Kansas and headed west. The Smoky Hill Trail quickly gained favor and praise from travelers and those that would benefit from the trade that travelers along the trail would provide—merchants, wagon makers and the authors of travel guides. As early as 1859, guidebooks and newspapers in Leavenworth, Junction City and Lawrence all touted the benefits of the Smoky Hill Trail in reaching the gold fields quickly and safely. One dissenting voice chimed in that "there was no discernible trail at all after one left Fort Riley…Added to this lack of knowledge of the route to be taken, those who recommended the Smoky Hill trail had little knowledge of distance" and said that "although it was the most direct, the Smoky was, due to scarcity of water, the hardest and most dangerous of the three great prairie roads from the Big Muddy to the Pike's Peak Gold Region."[100] Daniel Blue, his brothers Charles and Alexander and several friends had not read the dire warnings circulating

One of the most enigmatic figures in Kansas in the nineteenth century was General George Armstrong Custer. This *Harper's Weekly* engraving shows Custer discovering the remains of Lyman Kidder at the famed Kidder Massacre. *Courtesy Deb Bisel, author's collection.*

about the lack of safety and provisioning along the Smoky Hill Trail. In February 1859, the intrepid group left Lawrence, stopped at Fort Riley and then proceeded west for Pike's Peak. Taking the Smoky Hill Trail, of which they had no knowledge, the group became horribly lost. With provisions almost exhausted, the group made a pact that those who perished would become sustenance for those that remained. By the time Daniel Blue finally reached Denver, he was the lone survivor of a group of four that had pressed on in search of golden dreams and instead found death and desolation. For a time, travel on the Smoky Hill declined.

Human memory is fickle. How easily people forget the hardship and danger faced by those who have come before. By the mid-1860s, travel and commerce once again flowed along the Smoky Hill Trail. Utilizing a series of station stops, the Butterfield Overland Dispatch (BOD) transported mail, cargo and people westward through Kansas. The BOD transacted a huge amount of business along the Smoky Hill trail. On one day in July 1865, nineteen wagonloads of freight were received in Atchison and forwarded along the trail. Native Americans and thieves quickly realized the trail was a prime opportunity for raiding activities.

Located 50 miles west of Fort Hays and 180 miles from Fort Riley, Fort Downer was established along the Smoky Hill Trail in 1863 (near present-day WaKeeney, Kansas) initially as a station stop for the Butterfield Overland Dispatch. Fort Downer's heyday as a military post began in 1867. Along with Downer, Forts Hays, Monument and Wallace provided protection along the Smoky Hill Trail. With railroad building and settlers staking claims, tensions with Native Americans, particularly the Cheyenne, reached a fever pitch in the mid- to late 1860s. Fort Downer was no exception. The *Western Kansas World* noted, "Here in 1866 occurred the Fort Downer Massacre, in which all but one man were killed. Here Custer was encamped, and from this point and Fort Hays made several raids upon the wary red-skins."[101]

Fort Downer, like many of its counterparts built during the Plains Indian Wars, was transient in nature and short lived. Burned in 1867 during a raid, the post limped along until its closure and abandonment on May 28, 1868. Today nothing remains of the post.

Fort Bissell: 1872–1878

Following the American Civil War, so many veterans came to Kansas that it was dubbed "The Soldier State." Perhaps that bloody struggle was sound preparation for homesteading in the West.

When Phillips County and its county seat, Phillipsburg, were established in 1872, Indian raids were still a very real threat. For protection, the citizens built Fort Bissell on John Bissell's farm. The crude fort was made of cottonwood logs, stood up with points at the top. This was not a military post but one that could be ably manned by the locals when needed.

The fort is now a part of a pioneer museum complex on Fort Bissell Avenue in Phillipsburg.

FORT HAYS (FORT FLETCHER): 1865–1889

Custer. Cody. Hickok. Miles. Sheridan. The Buffalo Soldiers. To look at what remains of Fort Hays (originally called Fort Fletcher) today, one would hardly believe that such an illustrious assemblage of men and women once walked the now peaceful, windswept parade ground. Appearances can be deceiving. A state historic site, today Fort Hays is composed of two officer's quarters, a guardhouse and a blockhouse that were once teeming with activity. Motion-activated mannequins and interactive museum displays bring the story of Fort Hays to life for visitors as they step back to a time when the dreams of a growing nation clashed headlong into the cultural traditions of Native Americans holding on to their very existence on the plains.

Founded October 11, 1865, Fort Fletcher, the forerunner to Fort Hays, was located five miles south of present-day Walker. Like its other frontier garrison companions, Fort Fletcher was charged with keeping the trails clear

Fort Hays, once home to General George Armstrong Custer and his wife, Libbie, is now a state historic site. The Custers spent the summer of 1870 hunting buffalo and camping at Hays. *Courtesy Michelle M. Martin.*

and commerce running smoothly in Kansas. Initial troops stationed at the post were Galvanized Yankees from the Companies F and G of the First U.S. Volunteer Infantry commanded by Lieutenant Colonel William Tamblyn, supplemented by the Thirteenth Missouri Volunteer Cavalry. The Butterfield Overland Dispatch was the mover and shaker in the freight hauling business in Kansas. If you wanted goods moved west along the Smoky Hill Trail, the BOD was your best bet to ensure they arrived safely. Depredations on the part of Cheyenne and Arapaho Indians made transport of goods difficult, despite Fort Fletcher's best attempts to secure the trail. When David Butterfield went bankrupt, goods stopped flowing and the post was no longer needed. Abandoned quickly after its inception, Fort Fletcher seemed doomed.

Wild Bill Hickok, Texas Jack Omohundro and Buffalo Bill Cody all spent time at Fort Hays. Omohundro, a Virginian, had served with General J.E.B. Stuart in the Confederate army. His friendship with these two men, both of whom were veterans of the Union army, demonstrated how often differences could be laid aside for a new life in the American West.

The Union Pacific Railroad would be the salvation of Fort Fletcher. In October 1866, the former post was reopened and moved one-fourth of a mile north to be closer to water. Renamed Fort Hays, the post seemed destined to prosper. In 1867, however, a flood nearly wiped the post out. With the rail line paralleling the Smoky Hill Trail five miles north of the post, the flood was fortuitous. A new Fort Hays was constructed fifteen miles east of the flooded location, closer to the railroad right of way. This ensured that Fort Hays would become a supply depot for the army to supply its western posts and a troop garrison as well. To this lonely outpost came General George Armstrong Custer and his pretty bride, Elizabeth, better known as Libbie. The Custers and their circle of friends would leave behind a portrait of life at Fort Hays that is charming and heartfelt.

Arriving at Fort Hays in the summer of 1870, Annie Gibson Roberts Yates quickly became enamored with life on post and was befriended by the

Custers as well as Captain Frederick Benteen and his wife. Annie, writing in June while en route to Fort Hays, noted, "I kept a watch for buffalo all morning but with the exception of some which looked as if they had been shot from the train I saw none. Arrived at Fort Hays at 10 a.m. it is a pretty little fort situated near a lovely creek."[102] Almost immediately, the other officers' wives on post embraced Annie. Sewing, reading, horseback riding with male escorts and picnics in the area surrounding the post were pastimes of the officers' wives. Annie discovered that a military post had a pace all its own. "Sunday at a military post marches along pretty much as any other day. Then sun seems to shine warmer."[103]

General George Armstrong Custer spent time at various Kansas forts including Larned, Leavenworth, Hays, Harker, Riley and Wallace. *Courtesy Deb Bisel, author's collection.*

For Annie and the ladies on post, an outing into the prairie was a rare treat. On June 25, 1870, Annie ventured out into the unknown for a buffalo hunt. "We left at half past four in the ambulance—our horses being led. Gen'l. Custer & staff with an escort of 30 or 40 men. I rode in an ambulance with Gen'l. Schofield. I saw lots of antelope & prairie dogs & some rattlesnakes and when we had been out about 15 miles came in sight of three buffalo. Was a little scared but soon got over it...Afterwards saw a larger herd—Shot 8 of those. I retired early—my eyelids had ten pound weights on each one."[104] On a camping expedition away from post with the Custers and Captain George Yates, Annie experienced the thrill of the prairie and the dangers that came with it. On July 6, 1870, she noted in her diary, "In the early morning is the time Indians attack so we were fully certain a party of five or six scouts of our own were the enemy & more so when Genl. Custer and five or six officer ran over & charged into them & five or six shots took place."[105] She and the rest of the party were relieved when they received word that no Indian attack was imminent. However, her fear was palpable and justifiable as settlers lived a precarious existence during the time period. That afternoon, Annie would go riding with Captain

Above: The restored officer's quarters at Fort Hays. Rank brought with it some privilege on a military post. The quarters were duplexes that provided an officer and family ample room. Single officers shared quarters. *Courtesy Michelle M. Martin.*

Below: The parlor of one of the restored officers' quarters at Fort Hays. Social life on post centered around entertaining, and the parlor would have hosted afternoon tea and evening musical programs. *Courtesy Michelle M. Martin.*

Yates, whom she would later marry and take to the western trails with as an army wife. A frontier military post could also be the ultimate matchmaker as women were in short supply and men aplenty. The next day on the grand buffalo hunt, Annie reported seeing over ten thousand buffalo running over the Kansas prairie. Taking gun in hand, the delicate Annie felled a buffalo and brought "him to his knees & then to his death. We sang songs around another campfire."[106]

Much like the days of the buffalo roaming unfettered on the plains, the days of Fort Hays would also be numbered. As treaties with tribes were negotiated and they were relocated to reservations farther away from populated areas, the need for military outposts to protect trails and homesteaders waned. With the expansion of the rail lines in the west, goods and passenger trade could be conducted quicker and safer by rail, and the days of the stagecoach and Conestoga wagon on the trails also drew to a close. Fort Hays's closure in November 1889 signaled the end of the peopling of the great western frontier. Today, reminders of that storied past can be found in the restored officers' quarters, blockhouse and guardhouse of Fort Hays State Historic Site.

FORT DODGE: 1865–1889

Fort Dodge superintendent Carolos Diaz hears noises in the Custer House, he told a reporter. He tells himself it is just the wind.[107]

George Armstrong Custer did not have to be stationed at Fort Dodge in order to have a building named for him. Such is his fame. All he had to do was spend the night. Maybe it was a couple of nights. He arrived at Fort Dodge on October 9, 1868. His boss, General Phil Sheridan, was making his headquarters at the post. General William T. Sherman would have visited the post as well. The Seventh Cavalry camped nearby at Camp Sandy Forsyth. Prior to this, fighting was done in the summer, but officers and the public were growing weary of the "Indian problem" and the decision was made to attack during the winter when Indians were camped. When Custer and his men left Fort Dodge, they eventually wound their way into the Indian Territory for the Battle of the Washita. This action resulted in the death of Black Kettle, the Cheyenne leader who had survived the Sand Creek Massacre in 1864.

This was another post in that long line of defense on the Santa Fe Trail, being about midway between Fort Leavenworth and Santa Fe. Due to a

Left: The Kiowa chief Satanta was present at the Medicine Lodge Treaty Council in 1867 in Kansas. Captain Albert Barnitz wrote extensively about the chief in his letters and diaries. Today the town of Satanta, Kansas, in Haskell County is named for the chief. *Courtesy Deb Bisel, author's collection.*

Below: Referred to as the Custer House, this officer's quarters at Fort Dodge was a far cry from the dugouts officers lived in when the post first opened. *Courtesy Michelle M. Martin.*

scarcity of wood on the high prairie, the original 1865 structures were dugouts. Later, stone was used for constructing more durable buildings. Closed in 1882, the fort became a soldiers' home and remains so today. Its colorful history is the stuff of western legend. Historian Jeff Barnes described one of the more bizarre events. During the Hancock Expedition, led by the "Superb" Winfield Scott Hancock, the general met with the Kiowa chief Satanta at Fort Larned and honored the chief with the gift of a general's uniform. Days later, Satanta was proudly wearing the United States officer's garb when he and his band stampeded the horse herd at Fort Dodge.[108]

FORT WALLACE (CAMP POND CREEK): 1865–1882

Fort Wallace was established due to the continual presence of Native Americans and the encroachment of white civilization into their traditional hunting grounds and homelands. On October 26, 1865, General William T. Sherman ordered a new post be established on the Butterfield Overland Dispatch route in Western Kansas. Located near Pond Creek Station, a well-known home stop on the BOD, Camp Pond Creek was founded. In 1866, the post was renamed Fort Wallace in honor of Brigadier General William Harney Lamb Wallace, who died at the Battle of Shiloh. Like many forts before, Fort Wallace at first was nothing more than a collection of dugouts, tents and lean-tos. Intended to house four companies of troops (four hundred men) Fort Wallace was garrisoned by far fewer. As with many other military posts of the nineteenth century, the soldiers stationed at Fort Wallace were primarily responsible for its construction. Harsh weather and the continual threat of attack propelled the men in their work. By December 1866, the initial set of barracks was complete. These must have seemed like paradise to men used to sleeping in tents and lean-tos, exposed to the elements. Upon his arrival to Fort Wallace in June 1867, Captain Albert Barnitz found the post to be sturdy. "Five large buildings (of stone) have been completed here, and others are under way. Those finished are the Sutler's Store, commissary building, one company's quarters, a citizens and officer's mess house. The officers all live in tents or board buildings at present."[109] He also described the local stone quarry where troops toiled for their building material. It would take several years for the post to be completed. More than forty structures would be completed of stone and wood frame construction. Fort Wallace—due to its size, the number of

Every fort has its beginning. Captain Albert Barnitz (seated center) and officers lounge at Fort Wallace in Western Kansas. In June 1867, Barnitz engaged the Cheyenne for three hours near the fort. *Courtesy Deb Bisel, author's collection.*

troops it could garrison and their pluck and determination— would be called the "fightin'est fort in the west."

From 1865 to 1868, Fort Wallace would play an integral role in the protection of the western trails and commerce that depended on smooth travel in the region. Soldiers at Fort Wallace existed in a countryside brimming with Native American activity. During its tenure, the post would face off against more Native American armed conflict than any other Kansas post due to its proximity to the Smoky Hill Trail and the Colorado border. To ensue that Fort Wallace had the manpower needed to carry out its mission, Company I of the famed Seventh Cavalry, under the command of Captain Miles W. Keogh, was posted to Fort Wallace in 1866. When he assumed command of the post, he had at his disposal 195 officers and men. By 1867, troop strength had fluctuated at Wallace, with units dispatched on patrol and sent to other duty stations. Captain Albert Barnitz, commanding Company G, Seventh Cavalry, along with men from Company I, engaged the Cheyenne on June 26, 1867, in what would become the largest fight between troopers from Fort Wallace and the Cheyenne. For three hours, Barnitz and his men traded salvos with the Cheyenne in what was described as a "desperate little fight…doubtless the most extensive engagement that has occurred for some time on these plains."[110] Suffering seven fatalities, Barnitz and his men fared well. The mutilation of his troopers that died in the battle outraged other soldiers and made fodder for the newspapers back east. A photograph of Sergeant Frederick Wyllyams, his body horribly mutilated and arrow riddled, enflamed feelings of anger toward the Cheyenne on the plains. This would not be the only event connected to Fort Wallace that would arouse anger and animosity on the plains.

The men were desperate. With little in the way of supplies and surrounded by the Cheyenne on all sides, the future looked grim for Major General George A. Forsyth and his men as they battled against Roman Nose and his Cheyenne Dog Soldiers on the little island in Delaware Creek off the Republican River in eastern Colorado Territory. Fearing total defeat, Forsyth scrawled a note and dispatched it to Fort Wallace in Western Kansas on September 19, 1868: "I sent you two messengers on the night of the 17th instant, informing you of my critical condition. I tried to send two more last night but they did not succeed in passing the Indian pickets and returned. If the others have not arrived then hasten at once to my assistance. I have eight badly wounded

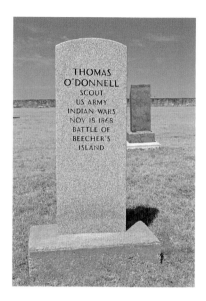

The grave of Thomas O'Donnell, one of Forsyth's Scouts from the storied Battle of Beecher's Island near Wray, Colorado. All that remains at the site of Fort Wallace is the lonely cemetery. *Courtesy Michelle M. Martin.*

This marker is all that remains of the post hospital at Fort Wallace. *Courtesy Michelle M. Martin.*

and ten slightly wounded men to take in and every animal I had was killed save seven which the Indians stampeded."[111] Fighting alongside Forsyth was young Lieutenant Frederick H. Beecher of the famed Beecher clan. He was dashing, gallant, handsome and charming. A veteran of the Civil War, Beecher made his way west and was posted to duty at Forts Riley and Wallace in western Kansas. After surviving savage combat in the war, Beecher could not have imagined the horror that would befall him when he "signed on to fight Indians" and engaged the much-feared Dog Soldiers. General George Armstrong Custer, in an 1872 article, described the hell of the Battle of Beecher's Island:

> *So close did the advance warriors of the attacking column come in the charge that several of their dead bodies now lay within a few feet of the entrenchments. The scouts had also suffered a heavy loss in this attack. The greatest and most irreparable was that of Lieutenant Beecher, who was mortally wounded, and died at sunset of that day. He was one of the most reliable and efficient officers doing duty on the Plains. Modest, energetic and ambitious in his profession, had he lived he undoubtedly would have had a brilliant future before him; and had opportunity such as is offered by a great war ever have occurred, Lieutenant Beecher would have without doubt achieved great distinction.*[112]

Buried where he was slain, Beecher gave his life in the competition between "civilization" and "savagery" on the plains. Fort Wallace, from its inception, was charged with safeguarding lives in the western Kansas frontier. Until its decommissioning in 1882, it more than lived up to that mission. Today, the fort itself is gone, and nature has reclaimed the land. Across from the location of the fort, the historic post cemetery stands a lonely sentinel on the landscape telling the story of Fort Wallace in stone. The Fort Wallace Museum in Wallace preserves the story of the post in word and artifact.

Monument Station: 1865–1868

Sometimes nature provides the best fortifications for man and spectacular scenery. Traveling in Kansas in 1865, Theodore H. Davis, a correspondent for *Harper's Monthly*, wrote colorful sketches of life in the rough-and-tumble West. Writing in 1865, he noted, "The Monuments were reached

Monument Rocks was another natural location that became a stage stop and gathering place for troops on the move along the Smoky Hill Trail. *Courtesy Michelle M. Martin.*

this evening; near them is a camp of more than 200 soldiers. A fort is to be built, also as a station. These Monument rocks are considered the most remarkable on the plains. At a distance it is difficult to realize they are not the handiwork of man, so perfectly do they resemble piles of masonry."[113] He painted a portrait of a post on the verge of civilization standing against the elements and enemies of man. "The wind that night was terrific. Two tents were blown away, and a wagon that was not brought into the corral overturned. The mules stood with their backs to the blast, that caused their hair to stand out like fur."[114] After a night like that, Davis was pleased to leave in the morning and was thankful for the presence of troopers at the station.

In the fall of 1865, Company A of the First U.S. Volunteer Cavalry and the Thirteenth Missouri Volunteer Cavalry arrived at Monument Station. Desolate, windswept and far from large settlements, the station was the perfect place for these troopers—they were Galvanized Yankees. Upon their release from Union prisons, these Confederate prisoners were shipped to far western Kansas, where they assuredly could not take up arms against the Union. Their services in keeping peace on the plains against Native American attack would allow Union troopers to continue the work of Reconstruction in the former Confederate States.

By 1866, Monument Station looked to be on the rise. Proposed as a full-fledged fort, the station limped along, suffering from lack of supplies and hoping for reinforcements and increased fortification efforts. Records indicate that in January 1866, supplies bound for Monument Station left Fort Fletcher. They never arrived, and the post was abandoned. In March 1866, infantrymen from companies A, E and I of the First Volunteer Infantry were sent to reoccupy Monument Station. Small in size, the station could only house one company, making it less effective as a duty station. Three buildings made up the fort complex at Monument Station. Existing drawings show a one-and-a-half-story limestone structure and two smaller outbuildings. The structures also served as a home and eating station on the Butterfield Overland Dispatch. After seeing limited military action, Monument Station was abandoned in June 1868. Today the Kansas prairie has reclaimed the land, and only natives to the area can make out the remnants of Monument Station.

Chapter 4

WAR: HOT, COLD AND BEYOND

1900–Present

I vividly remember my father telling me about seeing boxcar after boxcar filled with coffins passing through on their way from Fort Riley during the influenza outbreak during World War I.[115]
—*Betty Johnson Wallace*

For historian Frederick Jackson Turner, the frontier was "the meeting point between savagery and civilization." But according to the United States Census Bureau, by 1890 the American western frontier no longer existed. Peopled, developed, civilized, acculturated and built-up, the frontier—as earlier generations of Kansans had known it—had vanished. Along with the disappearance of the frontier came changes in the needs of the United States military in the West. Gone were the days of seeking out bushwhackers along the Missouri border and prolonged campaigns against Native Americans. Of the nineteenth-century posts built to protect trade, commerce and life in Kansas, only three—Fort Leavenworth, Fort Riley and Fort Dodge—remained at the turn of the century. New conflicts and challenges would require Kansas to become home to a new breed of military post. Protecting Americans in wars hot, cold and beyond, Kansas forts and bases are just as colorful, vibrant and historically significant today as their nineteenth-century forerunners.

General Dwight D. Eisenhower, supreme commander of the Allied Forces during World War II, and General George C. Marshall, chief of staff of the army from 1939 to 1945. Marshall appointed Eisenhower commander of the European Theater in 1942. Marshall, a graduate of Virginia Military Institute, graduated with honors from the Infantry-Cavalry School at Fort Leavenworth in 1907 and from the Army Staff College in 1908. Eisenhower, a West Point grad, finished first in his class at the Command and General Staff College in 1926. *Courtesy United States Army.*

CAMP FUNSTON: 1917–1925

It was all about numbers. Nearly forty thousand men trained at this World War I camp built next to Fort Riley in 1917; nearly four thousand buildings were constructed to serve them. Funston became a city unto itself. It had its own downtown, or "Zone of Camp Activities," with stores, libraries, schools, theaters, barbershops, bowling alleys, pool halls, fourteen YMCA buildings and three Knights of Columbus locations. At the end of the Great War, it was used to muster out the thousands of soldiers going home. Major General Leonard Wood was the camp's commander.[116]

The camp was named for one of the state's most famous sons, "Fighting Fred Funston" from Iola. The dapper officer served in the Philippines during the Spanish American War, was a commandant at Fort Leavenworth and was stationed in San Francisco during the 1906 earthquake.

Sadly, Camp Funston is remembered today for the worldwide flu pandemic said to have begun at the post. The "Spanish flu" was first reported in Haskell County, Kansas, and men at Camp Funston and Fort Riley soon exhibited signs. The first cases were reported at Funston in September 1918. By the end of the year, twelve thousand people had died in Kansas. A witness told the *Manhattan Mercury* that the soldiers were dying so fast "they were piling them up in a warehouse until they could get coffins for them." Wooden coffins with frozen bodies were "stacked like cord wood."

The problem of treating the ill was compounded by the fact that so many medical personnel were serving overseas. The flu spread around the globe through December 1920. Estimates of the dead ranged upward of sixty million people or about 3 percent of the world's population, far deadlier than the war itself.

The fate of Camp Funston itself was rather sad as well. Lucian K. Truscott Jr. was stationed at Fort Riley in 1925, and said, "In the eastern end of the reservation, on a broad flat between the escarpment and the Kansas River, Camp Funston had been the home of the Eighty-ninth Division, National Army, during the war. In 1925, the camp was completely demolished. Paved streets and foundation stones were all that remained, reminding one of a great cemetery, more so because of the Funston Monument standing at the entrance."[117]

KANSAS AIRFIELDS (ARMY AIRFIELDS AND NAVAL AIR STATIONS): 1941–1945

Even before the United States entered World War II, officials in Washington were keeping a wary eye on the developments in Europe. Should America be drawn into this war, or, if attacked, would she be ready? Could America defend herself? "Maybe" was the resounding answer. The nation began gearing up by the late 1930s, and then the unthinkable happened when America was attacked on December 7, 1941. The nation scrambled to build the machinery and train the people needed to win this war. It could be said that nowhere was the impact of the war effort more obvious than in Kansas. In November 1945, the *Kansas Historical Quarterly*, the publication of the state's historical society, carried the following notice: "Due to the absence of several members of the staff in war service, which makes it necessary for the other experienced members to take care of the routine demands on the Society, *The Kansas Historical Quarterly* for a time will be printed with fewer pages."

That same publication featured a summary by the staff of the contributions of Kansas to the war effort referred to as "The Battle of Kansas." In the first paragraph, that very Kansas attitude of humility is referenced.

> *During the war, most Kansans were so occupied with their specific jobs that they had little time to consider the state's tremendous contribution to victory. Two hundred thousand men and women went into the armed services. If the nation's casualty percentages are applied, there were 17,000 Kansas casualties, including 4,250 dead. At home on the farms, with less help than normal, the state's wheat yield for the four war years has not been exceeded by any similar period. Production of other crops and farm products was also high.*
>
> *Kansans helped build and staff the army and navy installations and the hundreds of war industries that dotted the state. Fort Riley expanded, with a new Camp Funston, and stressed mechanized cavalry. Fort Leavenworth continued its Command and General Staff School and also became a reception center. Several infantry regiments were trained at Camp Phillips, near Salina. The army located the 2,200-bed Winter General Hospital and an air force specialized depot at Topeka. Anhydrous ammonia was produced at the Jayhawk Ordnance Works near Baxter Springs, powder was manufactured at the Sunflower Ordnance Works near De Soto and shells were loaded at the Kansas Ordnance Plant near Parsons.*
>
> *Naval air stations were located at Olathe and Hutchinson. Army airfields were built near Salina, Topeka, Pratt, Walker, Herington, Great Bend, Liberal, Independence, Coffeyville, Dodge City, Garden City and Winfield. Varied types of training were given at these fields and from some, specially designated, departed thousands of the heavy bombers used in the European and Pacific war zones. Landing craft were built at Leavenworth and Kansas City and were floated to the gulf. Huge airplane factories were located at Kansas City and Wichita. At Wichita, the Boeing, Beech, Cessna and Culver factories completed 25,865 airplanes during the war and enough equivalent airplanes in spare parts to bring the number above 30,000. Boeing, Wichita's largest, employed as many as thirty thousand workers. This plant, under the management of Kansas-born J. Earl Schaefer, completed 8,584 Kaydet primary trainers and 1,762 additional*

trainers in spare parts; 750 CG4 gliders, the same gliders used in General Eisenhower's invasion of Europe; and wing panels and control surfaces for the B-17 Flying Fortress. Its work on the B-29 Superfortress was outstanding. All the B-29s used in the first raid on Japan on the steel center at Yawata on June 15, 1944, were built at Wichita and were processed from Kansas airfields.[118]

This was not just an honorific essay. Kansas State Historical Society staff George A. Root and Russell K. Hickman had witnessed history in the making, the sacrifices of accomplishments of their fellow Kansans. In no other issue of the *Kansas Historical Quarterly* is there such a proud and impassioned testament.

There were eighteen airfields in the state during World War II, sixteen army airfields and two naval air stations. Three were already in operation when the war began: Marshall Field at Fort Riley, Sherman Field at Fort Leavenworth and McConnell Field in Wichita. Construction began on twelve airfields in 1942 and one in 1943.

Fairfax Air Field was actually the municipal airport for Kansas City and was the precursor to the Olathe Naval Air Station. Used primarily to "ferry" planes from one field to another, the field had a lot of activity for a small installation. Women made big news at Fairfax in 1944:

A detachment of Women's Air Force Service Pilots (WASP's) was organized at Fairfax on 1 May 1944 to assist in the ferrying and did excellent service before being disbanded in September. Its head, Miss Helen Richie, held the woman's record for endurance flying, was the only woman to have served as a co-pilot on a commercial airline, and had been in charge of a detachment of American women transport pilots in England.[119]

Everything had to be done quickly, as evidenced by the Coffeyville Army Airfield where construction began June 1, 1942, and the post was activated a couple of weeks later, even though it was far from finished. By September, the staff could actually leave offices in downtown Coffeyville for space on the installation. This was the Army Air Force's Basic Flying School, and the mission was the basic, or second-stage, training of aviation cadets.

Many of the original personnel came from the Enid Airfield in Oklahoma. By February 1, 1943, there were 283 officers and 2,369 enlisted men stationed at Coffeyville. "From beginning to end, approximately 4,840 cadets and aviation students began the basic flying course, in 16 separate classes, at Coffeyville," reported authorities. "Incompletions, however, because of physical and flying deficiencies, serious accidents, and resignations were fairly numerous. As a result, only 3,881 successfully completed the course."[120]

Pilots for photo reconnaissance were also trained at Coffeyville:

> *During the 12-months' period ending on 4 June 1945 over 460 photo reconnaissance pilots completed all their training requirements at Coffeyville, and were shipped to staging areas for processing and assignment to overseas shipments. In addition, more than 200 pilots received their instrument flying training at Coffeyville, and were shipped to Will Rogers for training as photo reconnaissance pilots. There was no diminution in this indicated rate of training during the few remaining weeks of World War II.*[121]

Much of the construction for these airfields would have been the same for any community—streets, sewage system, water storage, gasoline and oil

Koss Construction was responsible for the majority of the army and naval airfield construction in Kansas. Koss Construction Company is seen here paving the first runways at Topeka Army Air Field in 1942. Note the air raid siren on the paver. *Courtesy Koss Construction archives.*

storage and electrical lines. Other projects, like runways and hangars, were evidence of the military mission.

Don Beuerlein, chairman of the board and past president of Koss Construction Company, recalled the experiences of his dad, Al Beuerlein, and uncle, John Beuerlein, who were construction superintendents for the company during the war. "We rarely saw Dad," he said. "They were pouring concrete in the rain, around the clock, no matter what the conditions. The water would ooze out of the construction forms as the concrete was poured in." These were not ideal conditions, but a necessity of wartime. Don recalled, too, that despite the incredible production schedule the company did not become rich with the effort. It was a conscious decision, he said, to be fair but not greedy in the face of the common threat to America.[122]

By 1941, Koss had already built roads and runways at Fort Leonard Wood in Missouri and Fort Smith in Arkansas. By April 1943, Koss had built thirty-four total miles of runway on various projects in those states and Kansas. An ad in the industry publication the *Construction Advisor* displayed a photo of the paving machine and crew in action with this caption:

A group of German World War II POWs gathers for a photo at the Concordia POW Camp. *Courtesy Deb Bisel, author's collection.*

While incarcerated, POWs looked for activities to fill their time. Writing and drawing gave the men a creative outlet. This drawing was a gift to the Schmanke family in Alma, Kansas. *Courtesy Deb Bisel, author's collection.*

Imagine a Runway 34 Miles Long!

Mister, that's a lot of runway…No matter how you figure it. It counts up to 3,000,000 square yards, and equals 620 acres. And it all has been built since December 7, 1941. Of course, it's not just one runway. But the total number of miles of runway we've built for the many great Army airbases located here in the middle-west. Naturally, we can't tell you their exact location or size. But someday…After victory…You'll be able not only to see them but actually use them. They are permanent…planned both for the present and the future. Then, and only then, will you thoroughly appreciate what a tremendous job was done in such a short span of time. Then, when you're using these runways and airfields, we would like you to remember this. The job could have been done in only one way…By free enterprise…By private contractors going all-out with their resources of men, money, machines, and engineering experience. And, after all, that's the real American way to do things![123]

It was the grueling schedule of World War II production that prepared the industry for "Ike's Road Plan" in the 1950s, the greatest public works project ever undertaken.

DODGE CITY ARMY AIRFIELD: 1942–1945

Young men and women from all over America, Canada, Great Britain and France came to this historic corner of Kansas to learn to fly. They named their bombers for Western heroes—Wyatt Earp, Bat Masterson, Bill Tilghman, Chalk Beeson, Brick Bond, Annie Rhule, Bob Wright, Calamity Jane and Annie Oakley. They named them for a legendary burying ground—Boot Hill Belle, Boot Hill Dust Storm, Boot Hill Buckaroo and Boot Hill Sodbuster. They named them for a distinctively western attitude—"Come and Get It."

Unlike his predecessors, Marshal Ham Bell lived to see his name emblazoned on the nose of a B-26 Marauder, and he had the honor of placing his handprint alongside it. It was July 4, 1943, and Bell was almost ninety years old. He was the longest-living Old West marshal, a legend within his own lifetime. It was often said that he never drew a gun in his years as a lawman, a fact that he clarified in the *Dodge City Daily Globe* of January 14, 1931:

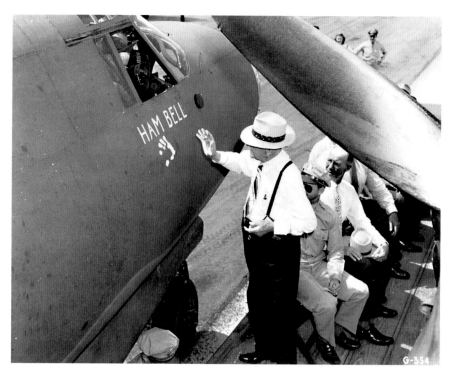

The last living "Old West" marshal Ham Bell christens the *Marauder*, named for him at the Dodge City Army Airfield. *Courtesy Kansas Heritage Center, Dodge City.*

Ham Bell says the idea that he never drew a gun on a man when he was sheriff here in the early days is all wrong. He never shot a man, he says, and that was mainly because he was always careful to draw his gun in plenty of time before the other man drew his. "If I'd never drawn a gun," he says, "I wouldn't have lived a week."

In 1943, World War II was raging. Kansas was at the heart of the nation and the heart of the war effort. Aviation and Kansas had become synonymous, and that same spirit of adventure that had characterized the Old West took to the skies. The high prairie landscape of Dodge City, certified as the windiest place in America, was a natural place to train the pilots desperately needed at the front. Plans were announced on June 11, 1942, and by September, the army had acquired more than 2,500 acres a couple miles from town and buildings were popping up faster than prairie dogs. Plans were to open in early 1943, and personnel poured into town, creating challenges for Dodge Citians. According to the 1944 *AAF Pilot School Yearbook*, "By December 7, first anniversary of Pearl Harbor, Dodge City had outgrown its residential properties. Thousands of new residents were in trailer houses; others were holed up in apartments that had been bedrooms; and many exclusive had been turned into boarding houses in preparation for the influx of personnel." Five days later, the Dodge City Army Air Field was officially activated.

The airfield created employment opportunities for civilians in various support capacities, such as chauffeurs. The presence of the base sometimes changed the routine in the quiet community, much to the curiosity of the locals. On March 1, 1943, the *Dodge City*

NOTICE TO THE AXIS:

NOW HE'S MAD!

Men Of The Air Force Welcome To Dodge

We're proud of Dodge City Army Air Field. We've come a long way in Dodge City in recent years . . . seen a lot of changes, but none in which we take greater pride than Dodge City Army Air Field, and the fine group of people it has brought to this community.

Yes, we are glad to have the Dodge City Army Air Field with us. And, we pledge our best effort to a continued support of the Field and its program, and to every effort towards winning the war.

Bishop Oil Company

Phone 191   222 S. Second

After the United States entered World War II, communities across the nation mobilized for war. Dodge City was home to an army airfield. This patriotic ad was designed to bolster community support for the war effort. *Courtesy Kansas Heritage Center, Dodge City.*

Globe reported that Fidelity State bank had not opened its doors for business one morning when military police armed with rifles showed up at each entrance. Military cars were "doubleparked" in front while bank personnel counted out the base's payroll. When the military personnel emerged with bags of money, everyone hurried into the waiting cars and were "swished away." It was a scene reminiscent of the Old West, where payrolls were to be guarded from outlaws.

With so many of America's able-bodied young men away at war, opportunity was created for women and minorities. The October 15, 1943 edition of the *Dodge City Globe* reported:

> *The first contingent of Texas tanned women pilots, wearing the silver wings of the 318th army air forces flying detachment arrived Thursday at the army airfield to train to fly the B-26 Marauder.*
>
> *The women, ranging from the wife of an AAF major to a co-ed who left college last April to turn aviatrix, are civilians with special civil service ratings enabling them to receive a strenuous nine weeks course here similar to the instruction given army air force pilots at this field...Pretty Miss Bonnie Jean Welz of Inglewood, Calif...was a student in the University of California at Los Angeles before she quit last April in her junior year to enlist in the flying training unit.*

When black soldiers arrived, the prejudices of the time were evident, though there were black pilots during World War II. On June 22, 1943, the *Dodge City Globe* stated:

> *The first group of Negro soldiers at the Dodge City army airfield has reached the field and is on duty now, it was announced Tuesday. The colored troops are from Midland, Tex., and they mostly are from home addresses in Texas and Arkansas.*
>
> *Their assignment is improvement of the field, landscaping, beautification and similar work in which they are engaging. They have separate quarters at the field, their own physical training program and operate as separate units.*
>
> *The colored populations of Dodge City, which is well organized socially with both men's and women's groups, are planning entertainments now for the newly arrived colored soldiers and John Schulte, USO director, said the USO is working on a program of recreation for colored soldiers similar to that of other troops.*

Then came the crashes. It is easy to forget that the whole purpose of the airfield was to *teach* flying, meaning the young pilots had to learn. With the urgency brought on by the war, time was a luxury, and men and machines were pushed to their limits. These headlines became all too common:

B-26 Crashes in Wheat Field Killing Crew
One Dead in B-26 Crash
Three Die in B-26 Crash
Three Die in B-26 Crash
One Dead Two Hurt in Crash
Marauder Crew Dead in Crash
Six Dead in Crash
Five Dead in Fall of a Marauder
Four Die in B-26 Crash
Five Died in Marauder Crash
Crews Die in Two Crashes

Often, the descriptions of the accidents, like this one from the *Globe* of September 13, 1944, were graphic and reflected the agricultural mindset of the community.

A B-26 Marauder from the army air field burst into flames and crashed shortly after taking off Thursday morning, in an unharvested wheat field on the Robert Pennington farm, eight miles west of Dodge city on the Beeson road, with one soldier burned to death in the plane…The plane plowed a deep furrow through the wheat field about 100 yards long as it crash landed, burning a swath of wheat in its path, but the wheat was damp enough that the rest of the field did not catch fire.

Many of the dead had moved wives and children to Dodge and had become part of the community. One of the more bizarre and tragic stories during this time was the death of Corporal Russel Stone of Weatherly, Pennsylvania. The young saxophone player in the post band had wanted to fly and had been denied. Finally, after months at the Dodge City Airfield, on September 8, 1944, he was cleared to go up and issued a parachute. He got onboard the plane, headed to the back and was not seen again. The parachute and his flight jacket were found next to an open camera hatch. Speculation ran from airsickness to suicide but there were not facts to substantiate either. His body was found in a pasture on March 2 the following year.

Nearly three thousand pilots were trained at the Dodge City Army Air Field. The August 1, 1945 *Dodge City Globe* reported a handful of soldiers and a maintenance crew by that time. In 1947, the field was sold at public auction. It is now in private hands. Cattle feed around the foundations and smoke stacks of a once vital military installation.

FORBES FIELD (TOPEKA ARMY AIR FIELD): 1942–PRESENT

Like most of the other airfields in Kansas, there was nowhere to house many of the first soldiers to report to duty in the summer of 1942. Fairgrounds were employed all over the state, and Shawnee County's agricultural hall was no different. Construction pressed forward, however, and by September, heavy bombardment training was underway. Crews were to receive thirty days of intense training before being shipped overseas. The mission evolved. The summer of 1943 saw the base processing and equipping B-17s and B-29s rather than training crews. According to an air force historian, one of the crews that passed through was headed by Colonel Paul W. Tibbets Jr., who piloted the B-29 Enola Gay that dropped the first atomic bomb on Japan.

Again, the location of Kansas in the geographic center of the lower forty-eight made it the ideal location for the mid-continent stop for ATC's "Statesman" in December 1945.

> [This was] *a daily transcontinental flight carrying key military and diplomatic travelers between Washington, D.C., and Hamilton Field, California. During December the base also became a stop for the "Globester," which provided daily shuttle service between Washington, D.C., and San Francisco. And in May 1946 the base took over operation of the daily "Alamo" flight between San Antonio, Tex., and Washington, D. C. Thus the field at Topeka became a major air terminal.*[124]

Following World War II, various activities took place at the base, including "Operation Santa Claus," a project that evacuated amputees from army hospitals. In the winter of 1946–47, members of the Portuguese Air Force trained at the airfield in air-sea rescue operations in B-17s and C-54s.

The base was deactivated on October 31, 1947, and reactivated July 1, 1948. It was inactivated on October 14, 1949, and reopened February 1,

1951, due to the Korean War. During the 1950s, a twelve-thousand-foot runway was built to handle RB-47s.

Following World War II, a different kind of threat defined the era. The "Cold War" was essentially an arms buildup and standoff between the United States and the Soviet Union. The two former allies entered a deadly competition to be the dominant global power, a competition that escalated to frightening proportions. The world had seen the devastating results of the atom bomb; for the first time, mankind truly had the technology to destroy itself.

In 1958, Forbes Air Force Base was notified that it would support Atlas E missile sites to be built nearby in Valley Falls, Dover, Waverly, Osage City, Delia, Wamego, Overbrook, Holton and Bushong. Construction began the following year. These missiles were stored horizontally and then raised up to fire. The missile sites, referred to as "coffins," were completed, and missiles were delivered in 1961. It was a short-lived program, and Secretary of Defense Robert McNamara directed decommissioning the missile bases in 1964.

Forbes Air Force Base was closed in 1973, but in 1974, the city of Topeka formed an airport authority, which received ownership of the property in 1976.

Home to the 190th Air Refueling Wing, the airfield is still named Forbes, but the municipal airport and industrial park was renamed in June 2012 as Topeka Regional Airport and Business Center.

Marshall Field: 1921–Present

One of the oldest army airfields, it was named for Brigadier General Francis C. Marshall, the assistant chief of cavalry, who had died in a plane crash the year before, and not General George C. Marshall, as is often assumed. The man who would become commanding general of the air force was at Fort Riley in 1912, and as a young lieutenant, Henry Arnold attempted to direct artillery fire from an airplane. His description of using stove pipe for sights and other crude, makeshift tools highlight the rapidly changing technology he would witness in his lifetime.

Again, the central location of Kansas was significant as the airfield was designed as the perfect place for cross-country refueling. It remains an integral part of the operation of Fort Riley.

SACRIFICE

Major Danny Forbes. *Courtesy Combat Air Museum.*

On June 5, 1948, Mrs. Dan Forbes sent a telegram from California to her mother in Ohio: "Leaving Los Angeles tonight for Topeka. Husband killed in aircraft accident today. "

Hazel Forbes was already a widow when she met the handsome, outgoing Major Dan Forbes. He had quite the service record. He had served in Elliott Roosevelt's photographic unit in North Africa, and he had taken the first aerial reconnaissance photos of Japan. Hazel's first husband had been killed in Europe fighting with Patton, and after his death, she went to work in Iowa at an installation specializing in communications. Dan was stationed there. He had a smile that would melt ice. It melted her heart, and they were married.

Then the war ended. Lots of boys came home and went back to civilian jobs. But as the Cold War loomed, a new threat appeared on the horizon and Dan remained in the military. When the Bikini atomic bomb tests occurred, he was filming and those clips were sent to Washington. He became a test pilot, one of that rare breed that revels in taking risks.

Dan was testing a YB-49, the Northrup Flying Wing described as "the fastest and most powerful bomber in the world." The plane was called a "monster" and weighed forty-four tons, with a wingspan of 172 feet and an overload gross weight of 209,000 pounds. Its eight jet engines were capable of a thrust equaling thirty-two thousand horsepower. Top speed was reported at five hundred miles an hour.

When it crashed near Muroc, California, Dan was one of five crewmen on board. They had been in the air about an hour when a witness said the plane seemed to explode and then fell to the earth, cutting a swath through the desert sagebrush and greasewood. There was no distress call. The crash occurred about ten miles north of the Army Proving Grounds at Muroc in a dry lakebed. The plane was demolished. Debris was everywhere. (Also killed on this flight was Captain Glen Edwards, for whom Edwards Air Force Base is named.) Years later, men scouring the crash site would find a blue gemstone. When Mrs. Forbes saw it, she went to her jewelry box and produced a ring with a matching gem— their wedding rings. She donated the gem to the Combat Air Museum in Topeka, and it is displayed with other artifacts and information on the life and career of Major Dan Forbes.

On July 19, 1949, the Topeka Army Airfield was renamed Forbes Field in honor of the native son who had given his life in the service of his country. More than three thousand people attended the events that included a flyover and marching military bands. Among the speakers were Governor Frank Carlson, Topeka mayor Frank Warren, Senator Arthur Capper and Alf Landon. Dan's mother was presented with the Distinguished Flying Cross on her son's behalf.

It was a clear, sunny day, and those perfect runways beckoned to the blue skies beyond.[125]

McConnell Air Force Base (Wichita Air Force Base): 1941–Present

Ironically, this major air force installation began as a municipal airport, unlike many other bases built during this time that were converted to municipal airports after the war. Wichita, long associated with the air industry, pushed for an airport almost as soon as the Wright Brothers left the ground. The far-thinking city fathers began buying land in 1916 with the airport in mind.

The town proved its commitment to flight in 1924 when it hosted the National Air Congress. Nearly fifty planes competed and performed, drawing a crowd of 100,000 observers. The public relations achieved help motivate Wichita voters to get behind the construction of a real airport. Ground was finally broken in 1929. The Wichita Airport became home to the National Guard in 1941, and in 1942, the Army Air Force Material Center moved in. Following the war, the military left until 1951 when the air force moved back and the facility became the Wichita Air Force Base. The fact that the airport was adjacent to the Boeing manufacturing plant was an incentive since the company is a huge defense contractor.

Gradually, the Air Force pushed the "municipal" right out the door and paid the city of Wichita to build another facility. Like Forbes in Topeka, the Wichita base became a support facility for missile bases in 1960. There were eighteen Titan II sites constructed in the area, twenty to fifty miles from the base.

No part of Kansas is immune to tornadoes, but the preparedness of the staff and the substantial storm shelters at the base certainly saved lives on April 26, 1991. The base was devastated, with buildings totally destroyed and tens of millions in damage. Miraculously, there were no deaths.

The proud aviation history of Wichita was reflected in the name-change to McConnell Air Force Base in 1954. The "Flying McConnell Brothers" made headlines during World War II, first because of the novelty of the three sibling pilots, then because of their exploits and, finally, their tragedies. In July 1943, Lieutenant Thomas McConnell was killed at Guadalcanal when his plane crashed in the fog, and Captain Fred McConnell crashed in Kansas after a bombing mission in 1945. Only Lieutenant Colonel Edwin McConnell survived the war. The base was originally named in honor of his two brothers because of regulations prohibiting the naming of installations for living honorees. Following Edwin's death in 1997, the record was amended to include him.

PRISONER OF WAR CAMPS: 1941–1945

The remote and "isolated" location of Kansas made it not only ideal for training flyers during World War II, but it also lent itself to housing the thousands of prisoners of war being apprehended by Allied troops. The largest such installation was Camp Concordia, but for a few months, there was a "satellite" camp in Lawrence. The German prisoners left a picturesque mark on the community.

When it was determined that the University of Kansas should have a chapel, officials turned to the two hundred German prisoners housed in town, many of whom had been stone masons prior to the war. It was ironic that the shortage of men created by the war should also be alleviated by the war. In fact, all over Kansas, Germans were working in fields and farms to keep the state's economy going.

The chapel was dedicated in 1946, and many couples can thank those Germans for building a beautiful place for them to begin their lives together. Countless weddings have been performed in the chapel in the decades since its construction. The chapel has evolved into a non-denominational building.

Kansas had eighteen POW sites, including the Disciplinary Barracks at Fort Leavenworth where Japanese "enemy aliens" were detained. Camp Concordia, a base camp, had more than four thousand prisoners at its largest capacity. Fort Riley was also a base camp. Other branch camps included those in Cawker City, Council Grove, El Dorado, Elkhart, Eskridge, Hays, Hutchinson, Neodesha, Ottawa, Peabody, Salina, Wadsworth and Winter General Hospital (now the VA hospital in Topeka). There was an Italian POW Camp at Kansas City, Missouri, as well.

The POW Camp Concordia Preservation Society operates a museum in Guard House 20, about two miles north of Concordia where the camp was located. Most of the original buildings are gone; some have been destroyed, and some were used for homes in Concordia. There is a two-story reconstruction of a guardhouse, visible from the road.

SCHILLING AIR FORCE BASE (SMOKY HILL ARMY AIR FIELD): 1942–1965

It was at the Smoky Hills Air Force Base, southwest of Salina, that the young pilot Carl Fyler met his dream plane. From his biography:

As Carl pulled up to the main building at the base, he was greeted with the glorious vision of a shiny new B-17F complete with ball turret and the new paddle bladed propellers. Carl couldn't wait. He had been scheming in his mind about this moment ever since he went to flight school. He and his crew scrambled into the beautiful bird and took her out for the test ride of her life. He headed straight out to western Kansas, with the clear blue skies and white clouds surrounding the plane. He made a beeline to Hodgeman County where his Uncle Joe Dvorak's ranch was located. As he buzzed over the roof of the house, he changed the prop pitch on all four propellers, which added even more decibels to the loud rumble the B-17 made normally, shaking the house and the ground for miles around. Uncle Joe ran out of the house. Aunt Lyda, who was hanging out laundry, dropped her basket, then realizing it was friend and not foe, grabbed a pillowcase and began waving it joyously at her nephew as he returned her greeting from the cockpit. Then Carl flew to Spearville. He directed the plane down the corner of Main Street. As he approached the steeple of the Catholic Church he did a maneuver called a chandelle up over the steeple. Aunt Lyda must have phoned ahead, because people ran out into Main Street waving towels. Then it was on to Hutchinson to give his parents a show too. Carl threw his head back and laughed in joy. He felt young and free, and he so loved to fly. For a while that afternoon, the thrill of flight overtook any concerns about heading to war.[126]

Fyler's experience reflected the adventure that lured young men and women not only to the military but also to the air. The relationship between Kansas and the sky became an unbreakable bond.

The Smoky Hills Air Force Base was renamed in 1948 to honor Colonel David Schilling, an ace with twenty-three kills against the German Luftwaffe. The Smoky Hills is also home to the Smoky Hills Weapon Range, the largest and most active Air National Guard range in the nation.

Walker Army Air Base near Hays was a satellite of the Smoky Hill Army Air Field and like Great Bend and Pratt, its runways were long enough and thick enough to handle the tremendous bombers built for the war. The B-29 Superfortress was built by Boeing. The company produced an incredible three thousand of the planes during the war. Weighing in at more than fifty tons and ninety-nine feet long, with a wingspan of 141 feet, it was amazing that the beast left the ground. Its range was nearly six thousand miles, and it was armed with twelve .50-caliber machine guns, one twenty-millimeter cannon and a twenty-thousand-pound bomb load.

One of the most tragic events involving the Superfortress occurred in Kansas in September 1944. The *Dodge City Globe* of September 18 reported:

12 Dead As A Bomber Hits Home
COPELAND—A giant, 4-motor bomber, from Walker army air base near Hays, crashed into the home and barn of the O. H. Hatfield home here late Sunday night, killing the ten man crew, Mr. Hatfield and his infant grandson, Jay Settles, and seriously wounding Mrs. Hatfield and her daughter, Mrs. Dean Settles.

In a Dodge City hospital Monday it was reported that Mrs. Hatfield and Mrs. Settles, though seriously burned and wounded with lacerations and bruises are expected to recover...

The Hatfield home, a half mile south of Copeland, was burned to the ground, the barn and various equipment were burned in a few minutes after the flaming bomber crashed.

The bomber, flying low and obviously in distress, burst into flames in the air over Copeland veered down hitting the Hatfield barn and then the house, both of which burst into flames as they were crushed down upon the four occupants.

Kansans paid a huge price to defend the world.

NOTES

Chapter 1

1. Myers, "From the Crack Post," 14.
2. Gower, "Pike's Peak Gold Rush."
3. Hallock, "Seige of Fort Atkinson."
4. Frazer, *Forts of the West*, 50–51.
5. Letter from John Brown to Orson Day, December 14, 1855. Kansas State Historical Society, available at www.territorialkansasonline.org/.
6. *Report of the Special Committee to Investigate the Troubles in Kansas*. 1193–99.
7. Ibid.
8. Hinton, *John Brown and His Men*, 398.
9. Preston, untitled article, *Lexington Virginia Gazette*, December 15, 1859.
10. Hoffhaus, "Fort de Cavagnial," 429.
11. Partin, *History of Fort Leavenworth*, 6.
12. Ibid.
13. Hoffhaus, "Fort de Cavagnial," 443–4.
14. Ibid., 448–9.
15. J.E.B. Stuart Papers. Virginia State Historical Society.
16. Davis, "Report of the Secretary of War," 119–20.
17. Martin, *Kansas State Historical Society 1903–1904*, 485.
18. Goodrich, "Somewhere Along the Solomon," 8.
19. Martin, *Kansas State Historical Society 1903–1904*, 497.
20. J.E.B. Stuart Papers.
21. Martin, *Kansas State Historical Society 1903–1904*, 500.
22. J.E.B. Stuart Papers.
23. Blackmar, *Kansas: A Cyclopedia*, 663.
24. Inman, *The Old Santa Fe Trail*, 111.

25. Ibid.

26. Chittenden, *American Fur Trade*, 588–9.

27. Walker, "Freighting: A Big Business on the Santa Fe Trail," 24.

28. Frazer, *Forts of the West*, 55.

29. Root, "Diary of Captain Wolf," 204.

30. Hunnius, Fort Larned National Historic Site. http://www.nps.gov/fols/ historyculture/pvt-adolph-hunnius.htm.

31. *Report of the Commissioner of Indian Affairs*, 571.

32. Utley, *Life in Custer's Cavalry*, 110.

33. Ibid., 177.

34. Partin, *History of Fort Leavenworth*, 20.

35. Ibid., 15.

36. Judge Adv. General Report, War Dept, 136.

37. Partin, *History of Fort Leavenworth*, 39.

38. Bucker, Bill. Personal correspondence, November 2012 with Debra Bisel.

39. Ibid.

40. Holder, Lieutenant General Don (ret). Personal correspondence November 2012 with Debra Bisel.

41. Ibid.

42. Ibid.

43. Fort Leavenworth website.

44. Garrard, *Wah-to-Yah*, 331–2.

45. Mitchell, *Linn County, Kansas*, 18.

46. Letter from Sene Campbelle to James Montgomery.

47. Ibid.

48. McKale and Young, *Fort Riley*, 9.

49. Ibid., 14.

50. Ibid., 38.

51. Ibid., 53.

52. Ibid., 128–9.

53. Oliva, *Fort Scott*, 6.

54. Ibid., 18.

55. Myers, "From the Crack Post," 6.

56. Lindberg and Matthews, Matt (eds.). "I Thought This Place Doomed."

57. Ibid.

58. Battle of Fort Titus website.

59. Kraft, *Ned Wynkoop*, 94.

60. J.E.B. Stuart.

61. Welch, *Border Warfare in Southeastern Kansas*, 69–70.

62. Tomlinson, *Kansas in Eighteen Fifty-eight.*

Chapter 2

63. Lindberg and Matthews, "I Thought This Place Doomed."
64. National Park Service. "Sand Creek Massacre History and Culture." http://www.nps.gov/sand/historyculture/index.htm.
65. Barry, "Fort Aubrey," 188.
66. Ibid.
67. Ibid.
68. Byers, *Daily Rocky Mountain News* [Denver], January 22, 1866.
69. *Topeka Daily Commonwealth*, September 3, 1872.
70. Allison, *History of Cherokee County*, 414.
71. *Fort Scott Bulletin*, June 28, 1862, 2.
72. Oliva, *Fort Harker*, 25.
73. Ibid., 27.
74. Ibid., 29–30.
75. Ibid., 32.
76. Ibid., 69.
77. Cutler, *History of the State of Kansas*, 342.
78. Langsdorf and Richard, "Letter of Daniel R. Anthony," 386.
79. Giles, *Thirty Years in Topeka*, 302–3.
80. Ibid.
81. Blackmar, *Kansas: A Cyclopedia*, 715.
82. Ibid., 424.
83. *75 Years in Great Bend,1872-1947: A Pictorial Pageant in 50 Pictures*, 35.
84. *Special Orders No. 185*, October 6, 1869.

Chapter 3

85. *Kansas Daily Correspondent*, May 27, 1868.
86. Campbell, "Camp Beecher," 183.
87. *Leavenworth Times and Conservative*, June 3, 1869.
88. Crawford, *Kansas in the Sixties*, 291.
89. A.L. Runyon's letters, *Kansas Historical Quarterly*, 61.
90. Mattes, "Patrolling the Santa Fe Trail," 576.
91. Ibid., 579.
92. Dyer, *War of the Rebellion*, 1641.
93. Kansas National Guard Museum.
94. Ibid.
95. Blackmar, *Kansas: A Cyclopedia*, 720–6.
96. Kansas National Guard Museum.
97. "The Butterfield Trail," http://www.skyways.org/towns/RussellSprings/bttrfld.html.

98. Roe, *Army Letters*, 86–7.
99. Ibid., 87.
100. Gower, "Pike's Peak Gold Rush," 155–7.
101. Blackmar, *Kansas: A Cyclopedia*, 660.
102. Pohanka, *Summer on the Plains*, 34.
103. Ibid., 37.
104. Ibid., 39.
105. Ibid., 42.
106. Ibid., 43.
107. Legends, 9.
108. Barnes, *Great Plains Guide to Custer*, 78.
109. Utley, *Life in Custer's Cavalry*, 68.
110. Oliva, *Fort Wallace*, 57.
111. Custer, "Beecher's Island."
112. Ibid.
113. Davis, untitled article, *Harpers Monthly*, July 1867.
114. Ibid.

Chapter 4

115. Betty Johnson Wallace, interview with Debra Bisel.
116. "Cantonment Life Camp Funston Illustrated (pamphlet)," Kansas State Historical Society.
117. Truscott, *Twilight of the U.S. Cavalry*, 79.
118. "The Battle of Kansas," *Kansas Historical Quarterly*, 480–81.
119. McGinley, "Wings Over Kansas," 129–57.
120. Ibid., 134.
121. Ibid.
122. Interview with Don Beuerlein.
123. Koss Construction Company Archives.
124. McGinley, *Wings Over Kansas*, 129–57.
125. Major Dan Forbes File. Topeka, Kansas. Combat Air Museum Files.
126. Webb and Norlin, *The Carl Fyler Story*, 52–3.

BIBLIOGRAPHY

AFF Pilot School Yearbook. Dodge City Army Air Field, 1944. Kansas Heritage Center Collections.

Allison, Nathaniel Thompson. *History of Cherokee County, Kansas and Representative Citizens.* Chicago: Biographical Publishing Company, 1904.

"A.L. Runyan's Letters from the 19th Century Kansas Regiment." *Kansas Historical Quarterly* 9 (February 1940): 58–75.

Barnes, Jeff. *The Great Plains Guide to Custer: 85 Forts, Fights & Other Sites.* Mechanicsburg, PA: Stackpole Books, 2012.

Barry, Louise. "Fort Aubrey." *Kansas Historical Quarterly* 39, no. 2 (Summer 1973).

Battle of Fort Titus. http://www.lecomptonkansas.com/page/the-battle-of-fort-titus.

"The Battle of Kansas." *Kansas Historical Quarterly* 13 (November 1945): 481–84.

Beuerlein, Don. Personal interview with Debra Bisel.

Blackmar, Frank W., ed. *Kansas: A Cyclopedia of State History Embracing Events, Institutions, Industries, Counties, Towns, Prominent Persons, etc.* 2 vols. Chicago: Standard Publishing Company, 1912.

Brown, John. Letter from John Brown to Orson Day. December 15, 1855. Kansas State Historical Society. Territorial Kansas Online. http://www.territorialkansasonline.org/~imlskto/cgi-bin/index.php?SCREEN=show_transcript&document_id=102510SCREEN=show_location&submit=&search=&startsearchat=&searchfor=&printerfriendly=&county_id=14&topic_id=&document_id=102510&selected_keyword=.

Buckner Family Collection, William C. Buckner.

Byers, William. *Daily Rocky Mountain News.* January 22, 1866.

Campbell, Hortense Balderson. "Camp Beecher." *Kansas Historical Quarterly* 3, no. 2 (May 1934): 172–86.

Campbell, Sene. "Letter from Sene Campbell to James Montgomery." Kansas State Historical Society, Manuscripts Division.

Chittenden, Hiram Martin. *The American Fur Trade in the Far West*. New York: Frances P. Harper, 1902.

Combat Air Museum Files. Major Dan Forbes File (loose clippings, telegrams). Topeka, Kansas.

Crawford, Samuel J. *Kansas in the Sixties*. Chicago: A.C. McClurg, 1911.

Custer, George Armstong. "Beecher's Island: A Thrilling Story of American Heroism." *San Francisco Chronicle*, December 8, 1872.

Cutler, William G. *History of the State of Kansas*. Chicago: A.T. Andreas, 1883.

Davis, Jefferson. "Report of the Secretary of War." *Senate Executive Documents*, Session 34th Congress, 3rd Session, December 2, 1856.

Davis, Theodore. Untitled article. *Harper's Monthly*, July 1867.

Dodge City Daily Globe. January 14, 1931.

Dodge City Globe. June 22 and October 15, 1943.

Douglas, Mrs. Isadore. Letter from Mrs. Isadore Douglas to her mother. February 3, 1867. Fort Larned National Historic Site Library.

Dyer, Frederick H. *Compendium of the War of the Rebellion*. Des Moines, IA: Torch Press, 1908.

Fort Larned National Historic Site. "Private Adolph Hunnius." http://www.nps.gov/fols/historyculture/pvt-adolph-hunnius.htm.

Fort Leavenworth Website. www.usacac.mil/cac.

Fort Scott Bulletin. June 28, 1862.

Frazer, Robert W. *Forts of the West: Military Forts and Presidios Commonly Called Forts West of the Mississippi to 1898*. Norman: University of Oklahoma Press, 1965.

Garfield, Marvin H. "The Military Post as a Factor in the Frontier Defenses of Kansas." *Kansas Historical Quarterly* 1 (November 1931): 50–62.

Garrard, Lewis H. *Wah-to-Yah and the Taos Trail*. Cincinnati, OH: W.H. Derby & Co., 1850.

Giles, Fry W. *Thirty Years in Topeka*. Topeka, KS: Geo. W. Crane Publisher, 1886.

Goodrich, Debra (Bisel). "Somewhere Along the Solomon." *Kansas Journal of Military History* 1 (Spring 2005): 8.

Gower, Calvin. "The Pike's Peak Gold Rush and the Smoky Hill Route, 1859–1860." *Kansas Historical Quarterly* 25, no. 2 (Summer 1959).

Hallock, Charles. "The Siege of Fort Atkinson." *Harper's New Monthly Magazine*, October 1957.

Hinton, Richard Josiah. *John Brown and His Men: With Some Account of the Roads They Traveled to Reach Harper's Ferry*. New York: Funk and Wagnalls, 1894.

Hoffhaus, Charles E. "Fort de Cavagnial: Imperial France in Kansas, 1744–1764." *Kansas Historical Quarterly* 30 (Winter, 1964): 425–54.

Holder, Lieutenant General (retired) Don. Personal interview/correspondence with author, November 2012.

Inman, Colonel Henry. *The Old Santa Fe Trail: The Story of a Great Highway*. New York: MacMillian Company, 1888.

Judge Advocate General Report. War Dept. 1916.

Kansas Daily Tribune. May 27, 1868.

Koss Construction Company Archives, Topeka, Kansas.

Kraft, Louis. *Ned Wynkoop and the Lonely Road from Sand Creek.* Norman: University of Oklahoma Press, 2011.

Langsdorf, Edgar, and R.W. Richard. "Letters of Daniel R. Anthony 1857–1862, Part One, 1857." *Kansas Historical Quarterly* 24 (Spring 1958): 6–30.

Leavenworth Times and Conservative. June 3, 1869.

Lindberg, Kip, and Matt Matthews, eds. "I Thought This Place Doomed: The Civil War Diary of Emma Caroline Morley." Unpublished manuscript.

Manhattan (KS) Mercury. March 1, 1998.

Martin, George, ed. *Transactions of the Kansas State Historical Society 1903–1904; Together With Addresses at Annual Meetings, Miscellaneous Papers, and a Roster of Kansas for Fifty Years.* Topeka, KS: George A. Clark, State Printer, 1904.

Mattes, Merril J. "Patrolling the Santa Fe Trail: Reminiscences of John S. Kirwin." *Kansas Historical Quarterly* 21, no. 8:569–87.

McGinley, Joseph. "U.S. Army and Air Force Wings Over Kansas." *Kansas Historical Quarterly* 25 (Summer 1959): 129–57.

McKale, William, and William D. Young. *Fort Riley: Citadel of the Frontier West.* Topeka, Kansas: Kansas State Historical Society, 2003.

Mitchell, William Ansell. *Linn County, Kansas: A History.* Kansas City: Campbell-Gates, 1928.

Museum of the Kansas National Guard Archives.

Myers, Charlene Scott. "Custer House Draws Visitors to Fort Dodge." *Legend: Life in Southwest Kansas* 11 (Fall 2012): 8–9.

Myers, Harry C. "From the Crack Post of the Frontier: Letters of Thomas and Charlotte Swords." *Kansas History, A Journal of the Central Plains* 5, no. 3 (Autumn 1982).

National Park Service. "Sand Creek Massacre National Historic Site." http://www.nps.gov/sand/historyculture/index.htm.

Oliva, Leo E. *Fort Harker: Defending the Journey West.* Topeka, Kansas: Kansas State Historical Society, 1998.

———. *Fort Scott: Courage and Conflict on the Border.* Topeka, Kansas: Kansas State Historical Society, 2000.

———. *Fort Wallace: Sentinel on the Smoky Hill Trail.* Topeka, Kansas: Kansas State Historical Society, 1998.

Partin, John W. *A Brief History of Fort Leavenworth 1827–1983.* Fort Leavenworth, KS: Combat Studies Institute and U.S. Command and General Staff College, 1983.

Pohanka, Brian C., ed. *A Summer on the Plains with Custer's 7th Cavalry: The 1870 Diary of Annie Gibson Roberts.* Lynchburg, VA: Schroeder Publications, 2004.

Preston, John T.L. Untitled article. *Lexington Virginia Gazette,* December 15, 1859.

Report of the Commissioner of Indian Affairs. 1865.

Report of the Special Committee Appointed to Investigate the Troubles in Kansas; with the Views of the Minority of Said Committee. House Report No. 200, 34th Congress, 1st Sess. Washington: Cornelius Wendell, printer, 1856.

Roe, Frances Marie Antoinette. *Army Letters From an Officer's Wife.* New York: D. Appleton and Company, 1909.

Root, George A. "Extracts From the Diary of Captain Lambert Bowman Wolf." *Kansas Historical Quarterly* 1 (1931–1932).

Russell Springs, Kansas. http://www.skyways.org/towns/RussellSprings/bttrfld.html.

75 Years in Great Bend, 1872–1947: A Pictorial Pageant in 50 Pictures. Great Bend, KS: N.p., 1947.

Sheridan, General Phillip. "Letter from General Phillip Sheridan to Governor Samuel J. Crawford." April 14, 1868. Kansas State Historical Society Manuscripts Division.

Special Orders No. 185.

Stuart, J.E.B. Papers. Virginia State Historical Society, Manuscripts.

Tomlinson, William P. *Kansas in Eighteen Fifty-Eight: Being Chiefly a History of the Recent Troubles in the Territory.* New York: H. Dayton, 1859.

Topeka Daily Commonwealth. September 3, 1872.

Truscott, Lucian K., III, ed. *The Twilight of the U. S. Cavalry.* Lawrence: University of Kansas Press, 1989.

Utley, Robert M. *Life in Custer's Cavalry: Diaries and Letters of Albert and Jennie Barnitz, 1867–1868.* New Haven: Yale University Press, 1977.

Walker, Wyman. "Freighting a Big Business on the Santa Fe Trail." *Kansas Historical Quarterly* 1, no. 1 (November 1931): 17–27.

Wallace, Betty Johnson. Interview with Douglass Wallace, January, 2005 by Debra Bisel.

Webb, Karl, and Ann Norlin. *11-11: The Carl Fyler Story.* BookLocker.com, 2012.

Welch, G. Murlin. *Border Warfare in Southeastern Kansas 1856–1858.* Shawnee Mission, Kansas: Fowler Printing, 2004.

Wings Over Kansas. www.wingsoverkansas.com.

INDEX

ABOUT THE AUTHORS

This is Debra Goodrich Bisel's second book with The History Press. *The Civil War in Kansas: Ten Years of Turmoil* was released in 2012. Deb is a graduate of Washburn University in Topeka, Kansas, with a B.A. in history. She serves as president of the Shawnee County Historical Society and promotes the history of many sites, including the Topeka Cemetery. She serves on the selection committee for the Kansas Hall of Fame (located at the historic Great Overland Station). She has appeared in numerous documentaries and talk radio shows across the nation on subjects related to the Civil War. She is writing the dual biography of the Civil War's first ladies, Mary Lincoln and Varina Davis. In addition to contributing to several magazines and websites, she blogs for *Mother Earth News*. Deb is married to Kansas Music Hall of Fame inductee Gary Bisel, and they make their home in Topeka.

Michelle M. Martin is a Michigander by birth and a Kansan by choice. Earning her B.A. and M.A. degrees in history from Western Michigan University, Martin arrived in Kansas in 1997. A history professor, living history interpreter, author and photographer, Martin founded her own historical consulting company—Discovering History—and serves as a historical consultant for the television, film and print industries. Her work has been aired on the History Channel, PBS and the National Geographic Channel. She is a living history interpreter and has volunteered her time at Fort Scott National Historic Site, Fort Larned National Historic Site, Wilson's Creek National Battlefield and Mine Creek Battlefield State Historic Site (Kansas) to name just a few. Her photographs have been featured in magazines and newspapers in the Central Plains. Martin's passion is preserving the rich tapestry of the American historical narrative and sharing it with the public. When not researching, writing and helping preserve history Martin enjoys storm chasing, photography, traveling to historic sites, hiking and spending time with her cat Josie.